SAE EDGE™
RESEARCH REPORT

Unsettled Technology Areas in Autonomous Vehicle Test and Validation

Dr. Rahul Razdan
Advanced Mobility Institute, Florida Polytechnic University

SERIES DEVELOPMENT TEAM

Dr. Mustafa Ilhan Akbas, Dr. Arman Sargolzaei, Dr. Ala' Jamil Alnaser, Dr. Saleem Sahawneh, Dr. Suleiman Alsweiss, and Dr. Jorge Vargas from *Advanced Mobility Institute, Florida Polytechnic University*

CONTRIBUTORS

Dr. Avinash Balachandran, *Toyota Research Institute*
Charlie Cheng, *SAPTC*
Dr. Xavier Fernando, *Ryerson University*
Apurva Kalia, *Cadence Design Systems*
Nicholas Keel, *National Instruments*
Karthik Krishnan, *MSC Software*
Dr. Dirk Langer, *Continental*
Jeff Lumina, *Jabil Circuit*
Dr. Raivo Sell, *Taltech*
Satya Sreenivas, *IBM*
Dr. Joachim Taiber, *International Transportation Innovation Center (ITIC)*
David Zuby, *Insurance Institute for Highway Safety*

SAE INTERNATIONAL®
Warrendale, Pennsylvania, USA

About the Publisher

SAE International® is a global association of more than 138,000 engineers and related technical experts in the aerospace, automotive, and commercial-vehicle industries. Our core competencies are life-long learning and voluntary consensus standards development. Visit sae.org

SAE EDGE™ Research Report Disclaimer

SAE EDGE™ Research Reports focus on topics that are dynamic, in which knowledge is incomplete, and which have yet to be standardized. They represent the collective wisdom of a group of experts and serve as a practical guide to the reader in understanding unsettled subject matter. They are not meant to provide a recommended practice or protocol. The experts engaged have contributed their own thoughts and points of view and these are not the positions of the institutions or businesses with which they are affiliated. A professional writer has collectivized their input; there is no one contributor's perspective being advanced, but rather that of a community of practitioners. SAE EDGE™ Research Reports are the property of SAE International and SAE alone is responsible for their content.

About This Publication

SAE EDGE™ Research Reports provide state-of-the-art and state-of-industry examinations of the most significant topics in mobility engineering. SAE EDGE™ contributors are experts from research, academia, and industry who have come together to explore and define the most critical advancements, challenges and future direction in areas such as vehicle automation, unmanned aircraft, IoT and connectivity, cybersecurity, advanced propulsion, and advanced manufacturing.

Related Resources

SAE MOBILUS® Automated & Connected Knowledge Hub
https://saemobilus.sae.org/automated-connected/

SAE Team:

Frank Menchaca, Chief Product Officer
Michael Thompson, Director, Standards, Information and Research Publications
Courtney Howard, Editorial Director
Monica Nogueira, Acquisitions Director
Jill Leonard, Product Manager
William Kucinski, Content Editor

Printed in USA

ISSN 2640-3536
e-ISSN 2640-3544

To purchase bulk quantities, please contact: SAE Customer Service

E-mail: CustomerService@sae.org
Phone: 877-606-7323 (inside USA and Canada)
724-776-4970 (outside USA)
Fax: 724-776-0790

Visit the SAE International Bookstore at books.sae.org

About the Series Editor

Dr. Rahul Razdan is senior director of special projects at Florida Polytechnic University in Lakeland, Florida. In this role, he serves as advisor to the president of the university and helps to manage the Advanced Mobility Institute, an applied research center for the development and testing of autonomous vehicle-related technology. Named on more than 24 issued patents, Razdan earned a Doctorate (Ph.D.) in Computer Science from Harvard University and both a Master of Science in Computer Engineering and a Bachelor of Science in Electrical and Computer Engineering from Carnegie Mellon University.

Razdan boasts 35 years of experience in startups, academia, and Fortune 500 companies working in such areas as science, technology, engineering, and mathematics (STEM) education; autonomous vehicle technology; and semiconductor design. The Association for Computing Machinery (ACM) and the Institute of Electrical and Electronics Engineers (IEEE) have recognized him with "Hall of Fame" inductions. Razdan has also won numerous best-paper awards and has led several successful startups in areas such as electronic design automation (EDA), wireless power, artificial intelligence (AI), and machine learning (ML).

contents

Unsettled Technology Areas in Autonomous Vehicle Test and Validation

Abstract

Automated driving system (ADS) technology and ADS-enabled/operated vehicles - commonly referred to as automated vehicles and autonomous vehicles (AVs) - have the potential to impact the world as significantly as the internal combustion engine. Successful ADS technologies could fundamentally transform the automotive industry, civil planning, the energy sector, and more.

Rapid progress is being made in artificial intelligence (AI), which sits at the core of and forms the basis of ADS platforms. Consequently, autonomous capabilities such as those afforded by advanced driver assistance systems (ADAS) and other automation solutions are increasingly becoming available in the marketplace. To achieve highly or fully automated or autonomous capabilities, a major leap forward in the validation of these ADS technologies is required. Without this critical cog, helping to ensure the safety and reliability of these systems and platforms, the full capabilities of ADS technology will not be realized.

This paper explores the ADS validation challenge by reviewing existing approaches and examining the effectiveness of those approaches, presenting critical techniques required to bring safe and effective solutions to market, discussing unsettled topics, and suggesting next steps for industry stakeholders to consider as they work to advance the ADS ecosystem.

NOTE: SAE EDGE™ Research Reports are intended to identify and illuminate key issues in emerging, but still unsettled, technologies of interest to the mobility industry. The goal of SAE EDGE™ Research Reports is to stimulate discussion and work in the hope of promoting and speeding resolution of identified issues. SAE EDGE™ Research Reports are not intended to resolve the issues they identify or close any topic to further scrutiny.

DR. RAHUL RAZDAN
Advanced Mobility Institute
Florida Polytechnic University

Series Development Team

Dr. Mustafa Ilhan Akbas
Dr. Arman Sargolzaei
Dr. Ala' Jamil Alnaser
Dr. Saleem Sahawneh
Dr. Suleiman Alsweiss
Dr. Jorge Vargas, *Advanced Mobility Institute, Florida Polytechnic University*

Contributors

Dr. Avinash Balachandran, *Toyota Research Institute*
Charlie Cheng, *SAPTC*
Dr. Xavier Fernando, *Ryerson University*
Apurva Kalia, *Cadence Design Systems*
Nicholas Keel, *National Instruments*
Karthik Krishnan, *MSC Software*
Dr. Dirk Langer, *Continental*
Jeff Lumina, *Jabil Circuit*
Dr. Raivo Sell, *Taltech*
Satya Sreenivas, *IBM*
Dr. Joachim Taiber, *International Transportation Innovation Center (ITIC)*
David Zuby, *Insurance Institute for Highway Safety*

ISSN 2640-3536

Introduction

Automated or autonomous vehicle technology can be defined as the progressive shift in driving responsibilities from a human driver to an automated driving system (ADS). Indeed, a majority of vehicle manufacturers have been moving in this direction [1]. To understand the role of testing and validation in this shift, it is important to review the economic drivers, existing techniques, and current challenges.

Economic Drivers for Autonomous Vehicle Technology

Several high-level statistics, listed here, are relevant to understanding the potential impact of ADS technology.

- Average vehicle utilization in the United States is less than 4% (96% idle time) [2, 3].
- In 2017, more than 90% of the 40,000 vehicle fatalities - costing roughly $410 billion - were due to human error [4, 5].
- In the United States, 40% of Uber rides are under 2 miles, with 60% of the cost of fare associated with the vehicle driver [6].
- Average United States road and parking lot utilization is less than 10% [2, 3].
- The lack of access to transportation limits the engagement to civic life for nondriver groups, such as children and the elderly [7].

Given these statistics, the pervasive utilization of ADS technology will have a transformative effect on safety [8], travel behavior and time [9], traffic flow stability [10], fuel consumption [11], public health, and parking and infrastructure utilization [7, 12, 13].

Even the core concept, the need for personal vehicle ownership, is likely to change. In urban and suburban areas, vehicle ownership could be reduced via an automated fleet of on-demand service vehicles. Individually owned vehicles could also be brought into a commercially operated fleet when not in use by the owner, providing a source of income for both the owner and fleet operator.

Based on utilization, therefore, an order-of-magnitude fewer vehicles might be needed as compared to today, despite any short-term increase resulting from usage and access. The economic effect will enable disruptive new business models, which can raise productivity in massive markets, such as passenger transportation, construction (road, residential, and commercial), agriculture, and logistics.

State of the Industry

The SAE International J3016™ "Levels of Driving Automation" standard defines automated levels for vehicles as "Level 0 - No Automation," "Level 1 - Driver Assistance," "Level 2 - Partial Automation," "Level 3 - Conditional Automation," "Level 4 - High Automation," and "Level 5 - Full Automation" [14].

Level 0 through Level 2 automation capabilities are available in the marketplace and the functions at these levels require either full or partial human driver control or oversight to perform vehicle operation in the U.S. Although Level 3 is also available in the marketplace - e.g., the Audi A8 - regulators have not approved the utilization of Level 3

FIGURE 1.

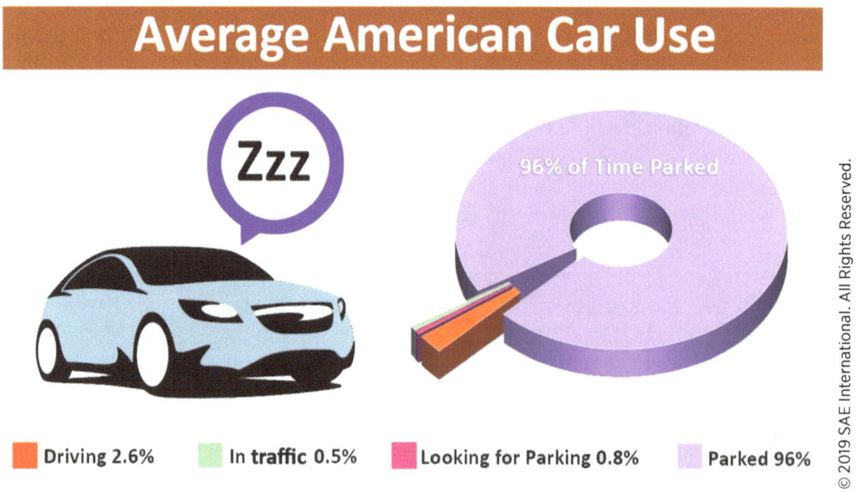

technology in the U.S. At Level 3, the role of driver is necessary to ensure human intervention when required. With Level 4 and Level 5 automation, the vehicle is capable of performing all driving functions, making driver control optional.

Vehicle communication capabilities and connectivity are also evolving. A variety of technologies have been developed to enable connected vehicles (CVs) and connected automated vehicles (CAVs). Advances continue in vehicle-to-vehicle (V2V), vehicle-to-infrastructure (V2I), vehicle-to-network (V2N), and vehicle-to-everything (V2X) connectivity and communications in the intelligent transportation system (ITS) [15, 16].

ADS technology has attracted actors and investment from different areas, including traditional automotive manufacturers or original equipment manufacturers (OEMs), high-technology organizations, Tier 2 and Tier 3 providers of AI and electro-optics/sensor technologies, data-centric technology and service providers, and hardware, software, and service providers from other market verticals. Technical standards organizations, regulators, and insurance companies are also important actors who aim to help shape the future of ADS-enabled vehicles.

Vehicle manufacturers are the most active group in this field to date. Conventional vehicle manufacturers, such as Ford Motor Company and Jaguar Land Rover, and electric vehicle companies, such as Tesla, are investing heavily in ADS research and development. Additionally, ADS-specialized companies, such as Waymo, and transportation network companies, such as Uber and Lyft, are pioneering the utilization of this technology. These companies have invested billions of dollars, and have been productive in practical applications, research, and development [17].

The most successful examples of ADS technology are still at Level 2, at best, despite significant investment from multiple players. Various reported accidents involved Level 1 or Level 2 systems, according to California's DMV AV Accident Report [18]. For instance, an autopilot feature was released too early in one of these cases, causing a fatal crash [19].

Level 2 capability has proven suspect in recent studies from the Insurance Institute for Highway Safety [20] where road curvature, tree shadows, and even stationary targets confused existing ADS-enabled vehicles.

Thorough ADS test and validation approaches are needed and will go a long way toward ensuring ADS and ADS-enabled vehicle safety and reliability. This SAE EDGE report considers the following five areas to be unsettled domains in ADS test and validation, as marked by an apparent lack of common understanding regarding key aspects of each:

- Replacing the driver, compressing human learning and understanding unwritten rules of the road
- Regulations, which inform and affect ADS test and validation approaches
- Finessing existing validation approaches to fit new ADS paradigm
- Hardware validation and natural language processing
- Progressive structures, evolving requirements to ensure future safety

FIGURE 2.

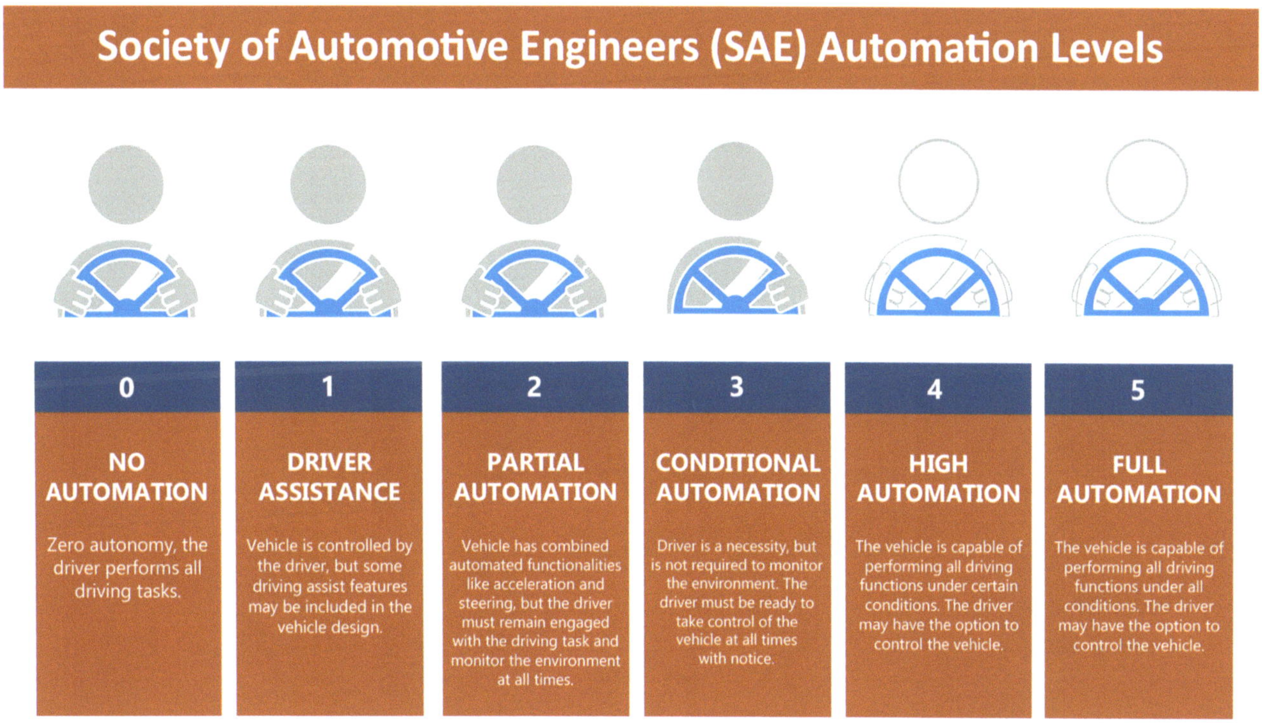

Replacing the Driver

The successful operation of a vehicle in today's infrastructure requires three major steps: perception, decision-making, and action. In a conventional vehicle, the perception and decision-making steps are controlled by the driver with the action step executed by the vehicle. The automotive industry has a long history in the validation of the action steps; ADS perception and decision-making are relatively new challenges. To have a robust ADS-enabled vehicle, we must answer the question: "How much time does it take for an autonomous system to replace the driver?"

Individuals begin noticing and potentially start learning the rules of the "human driving world" from as early as the first time they ride in the backseat of a vehicle as a child, and that learning experience continues over the next roughly two decades. It is during these years of operation in the human driving world that drivers learn how to perceive an environment with other humans and take nonlethal actions. Even though actual driver training courses on the tail end of this long training are useful, the number of "unwritten rules of the road" could potentially outweigh or outnumber those in the official rule book.

Many driving situations exist in which drivers must make a decision in response to an event for which they were not explicitly trained. The short answer to the question as to the time required for an ADS to replace the driver, therefore, would be the time equivalent to the number of years a human spends in learning both official and unwritten traffic rules.

From a validation point-of-view, this presents four significant unsettled topics, described as follows.

- Short development cycle: The long period of human learning must be compressed into a development cycle appropriate for a product (not 18 years).
- Unwritten traffic rules: An ADS must understand the dynamic unwritten rules of the traffic. As an example, a leaf or a pigeon is not an obstruction while a stone may be. Further, the ADS must be able to operate in an environment with humans and obey unwritten rules.
- Interference and fusion: The sensor systems must build an accurate version of reality, even in highly "noisy" environments such as rain, snow, or sleet. Further, the sensory systems (as well as the internal electronics) must be resistant to human-made and natural sources of electromagnetic noise or interference (e.g., lightning or transmission lines).
- Progressive structure: The validation structure must always move forward in terms of safety and reliability.

ADS are largely expected, by the general public, to behave in a manner consistent with a "reasonable human" given that ADS-enabled vehicles will operate within a mixed human-machine environment within current legal structures. This consideration raises additional questions and

FIGURE 3.

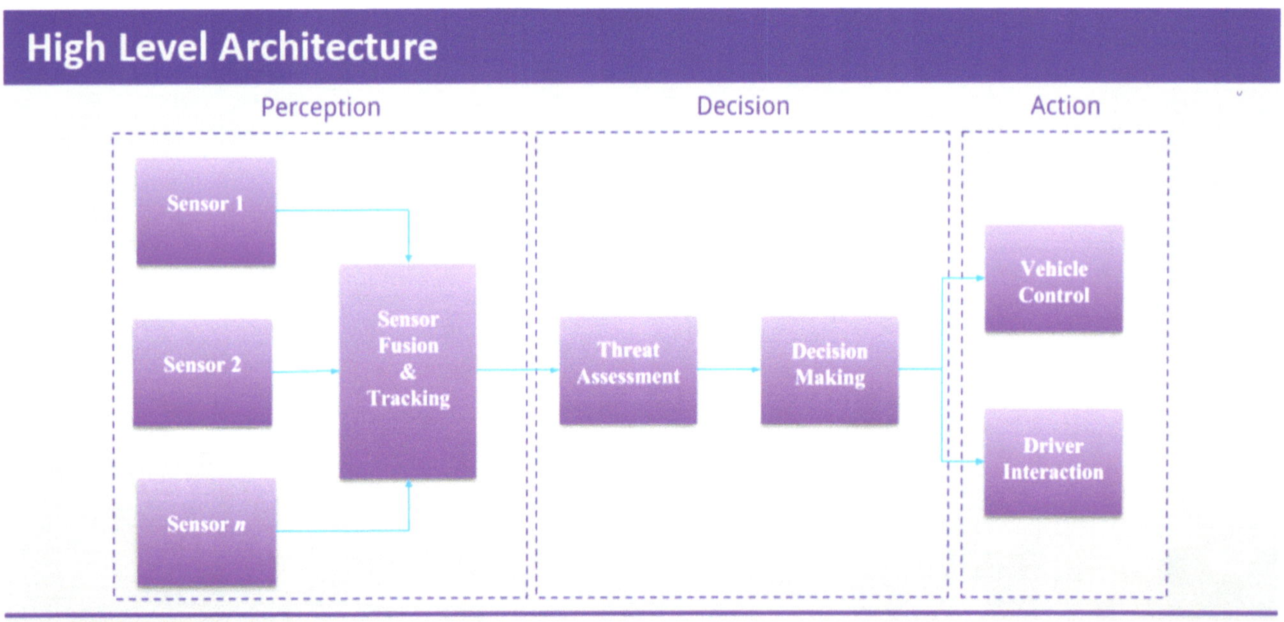

issues, perhaps around the objective, rather than subjective, defining characteristics of a "reasonable human" and ethics [21] - both potentially unsettled topics to be addressed another time. Security is another critical and unsettled domain which must be managed and will be addressed in another report.

For more advanced ADS systems, an analysis of the accidents to date [22] has shown that a majority of ADS-related collisions involve human-operated vehicles hitting ADS-enabled vehicles at a higher rate than human-operated vehicle colliding with another human-operated vehicle. The reason is that ADS platforms do not yet conform to the social norms and nonverbal communications that human drivers use. ADS adaptation to social norms and nonverbal communications, such as through machine learning (ML) technologies and capabilities, is an unsettled domain to be explored at another time.

Given the daunting challenges in open-field ADS systems, various restrictive or simplifying solutions have been proposed. Ideas such as dedicated lanes have been proposed for AVs or campus situations where AVs are isolated from other vehicular traffic. These simplifications certainly help with the verification of safety, but at significant cost for initial buildout and ongoing maintenance. This is in a context where the current infrastructure is already in dire need of maintenance upgrade. Finally, it is unclear when the number of AVs deployed will be sufficiently large to justify a dedicated lane; until that point, an obvious trade-off exists in providing resources for AVs as opposed to human drivers – creating a chicken-and-egg situation.

Further, AVs are quite a bit more expensive than conventional automobiles, driven by the cost of the electronic systems. The current cost is one of the key factors slowing mass deployment of AV systems. The method for reducing cost is well understood through the use of semiconductor integration. Semiconductor design is very expensive and only justified with large volumes; however, one cannot get to large volumes until costs are reduced. Thus, another chicken-and-egg situation.

Radio-frequency (RF) technology is, likely for the first time, being contemplated for use in the inner loop of consumer safety-critical systems. It marks a significant escalation in the need for robustness as compared to a cellular phone call or Wi-Fi hotspot. Electromagnetic interference (EMI) from natural (weather) and human-made (power plants, transmission lines) sources can threaten myriad AV systems. This interference can also be a function of the road topology (tunnels, bridges).

Mobileye, a subsidiary of Intel Corporation, unwittingly demonstrated the dangers of EMI in an incident whereby electronics used by journalists at an event produced sufficient EMI to interrupt dedicated short-range communications (DSRC) between the traffic light to the AV [23]. The next SAE EDGE Research Report in this series on test and validation addresses unsettled domains relative to EMI from natural and manmade sources and the effect on ADS.

ADS validation is not entirely new; rather, automated and autonomous air, space, and marine systems have previously undergone sometimes extensive test and validation, as examples. ADS-enabled ground vehicles are, however, an order of magnitude more difficult to build and verify as opposed to air, space, and marine ADS-enabled systems because of two considerations: density and human behavior prediction.

Compared to aerospace, ADS-enabled ground vehicles must operate only on paved surfaces with several orders of magnitude higher number of players in one less dimension. Additionally, ADS-enabled ground vehicles must constantly predict the behavior of other actors to avoid collision. The combination of these factors makes the validation of ADS-enabled ground vehicles more difficult.

Overall, validation methods are at the center point of many aspects of AV ecosystem. The sophistication and progress in the validation methods is one of the gates that determine the investment mix between automobile and infrastructure. The progress in validation is also a key aspect of building confidence with regulators and most importantly of building trust with the final consumer.

Recommendations

1. Language of Driving: A far better scientific understanding is required of the language of driving and the associated dialects in various regions used by human drivers. Research is required to understand this language and fold into an overall verification framework.
2. Electromagnetic Interference: Since many of the proposed solutions require the use of RF technology for safety, it is vitally important to understand the impact of sources of EMI from natural (weather) as well as human-made sources (power plants).
3. Progressive Structure: One cannot have a safety protocol that inspires confidence without an architecture that demonstrates a progressive improvement. Research is required on system architectures for safety that have this characteristic.
4. Infrastructure vs Self-standing: Solving technical shortcomings with the aid of infrastructure is possible. However, the additional investments must be economically viable. Research is required to study the situations in which these investments are prudent.

Regulations

With the massive shift of functionality from the driver to the ADS, regulatory issues for ADS-enabled vehicles need to be considered as technical plans for ADS technology are implemented [24, 25, 26]. The tremendous potential economic value for ADS technology has effectively created a race between states in the U.S. to enable ADS-enabled vehicle design, testing, and deployment.

Florida, Michigan, and Arizona have been in the forefront of enabling ADS testing after California's initial work [27]. On the federal level, the National Highway Traffic Safety Administration (NHTSA) produced two versions of its Preliminary Statement of Policy on Autonomous Vehicles and the U.S. Department of Transportation (DoT) issued its three reports on ADS policy [28] for providing safety assurance and facilitating innovation.

While regulating the driver has traditionally been a state function, the U.S. DoT policy outlines options for the further use of federal authorities to expedite the safe introduction of ADS-enabled vehicles into the marketplace. The number of states considering ADS legislation to eliminate fatalities and serious injuries has been steadily increasing [29]. According to the National Conference of State Legislatures' (NCSL's) Autonomous Vehicles Legislative Database, 33 states introduced legislation in 2017, whereas this number was 6 in 2012 [30]. Similarly, 41 states have been discussing legislation since 2012.

In Germany, ADS-related legislation was approved as an amendment to the German Road Traffic Act [31]. This amendment includes provisions on ADS-related concepts, such as liability and data storage. The regulatory changes are going to affect insurance companies and legal entities because liability is an important factor to consider with ADS technology. Liability considerations are critical for vehicle manufacturers as ADS-enabled vehicles have the potential to shift responsibility to the manufacturer. Liability is another unsettled domain to be addressed at another time in an SAE EDGE Research Report.

Another pressing regulatory challenge is the lack of a technical language for communication between regulators and vehicle manufacturers or technology providers on safety and testing issues. An emerging need exists for a validation framework that will fill this gap.

Recommendations

An efficient validation framework is needed and must provide a platform that includes test scenario generation and scenario coverage for these tests. When an accident occurs, regulators must be able to easily build the failing scenario and add it to the coverage stack. The overall environment must be designed to make positive progress in terms of safety. Similarly, this framework must provide a method for industry to communicate and build a shared test set for mutual use for safety and validation. Such a platform would provide the capability of certification for ADS technology and creation of safety regulations using a common language between regulators and industry.

1. Technical Interface Standards: The need exists for technical standards that expose the inner workings of key aspects of the AV. Critical information includes: what the AV saw as a result of the perception stage, what communications were broadcast and received by the AV, and the AV's calculation for threat analysis - all of which are likely to be at the center of an AV accident investigation, in lieu of the typical driver interview done routinely by law enforcement. Note: The actual algorithms for perception, decision-making, or threat analysis can be proprietary.
2. Management of Systemic Risk: Unlike human drivers, AVs have a much higher systemic risk

FIGURE 4.

because large fleets are running the same software. Thus, to manage this risk, a much higher sensitivity must be afforded for the detection and correction of systemic issues. In other industries, such as airlines in the aviation market, it is accomplished by providing safe and anonymous methods to report abhorrent behavior.

3. Perception by humans: Accidents can be caused by humans' perception or anticipation of an AV's future behavior; therefore, significant deviations from "normal" or expected behavior must be captured and corrected.
4. Learn from other industries: DoT and NHTSA guidance documents have recognized the need for an upgraded infrastructure, urging the AV industry to learn from other industries. Significant updates based on aerospace industry concepts such as the black box, vehicle accident report, accident databases make good sense. The NHTSA's *Automated Driving Systems: A Vision for Safety* reads: "Entities are encouraged to adopt voluntary guidance, best practices, design principles, and standards developed by established and accredited standards-developing organizations (as applicable) such as the International Standards Organization (ISO) and SAE International, as well as standards and processes available from other industries such as aviation, space, and the military and other applicable standards or internal company processes as they are relevant and applicable." This direction is prudent, but the AV task is quite a bit more difficult.

Finessing Existing Validation Approaches to Fit New ADS Paradigm

Traditionally, complicated systems such as automobiles have extensively used the V paradigm at both the system and component levels. In this paradigm, system models of behavior are successively refined in the process of synthesis of the design and successively verified at the various levels with the physical realization of the systems.

In terms of the three major steps in ADS-enabled vehicles - perception, decision-making, and action - the automotive industry has an established process for the design and validation of the action step, based on more than 100 years of experience. With the introduction of ADS-enabled vehicles, the perception and decision-making steps have redefined the system in a fundamental manner. Several approaches have been attempted to address the challenges of ADS and ADS-enabled vehicle validation. These approaches focus on either one subsystem or a combination of ADS subsystems. The most important existing validation approaches can be grouped as follows.

1. Test tracks
2. Real-world driving
3. Video and image datasets
4. Modeling and simulation

The current state of these approaches and their advantages and disadvantages are presented in the following sections.

FIGURE 5.

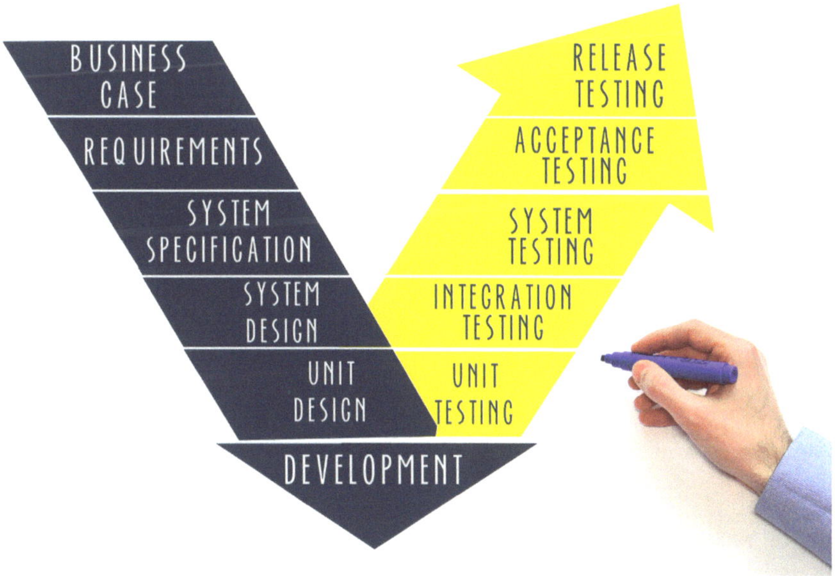

Test Tracks: Advantages and Disadvantages

Test tracks have been built and utilized all around the world as given in Figure 6. Some recent tracks have been built solely for ADS testing, others have recently started to be used for ADS testing.

These tracks are large establishments, which include some or all of the following facilities and features: vehicle installation and maintenance facilities, fueling facilities, tracking throughout site, traffic operations centers, demonstration and observation facilities, ITS infrastructure, site security, global positioning system (GPS) mapping, cybersecurity and privacy, conference and meeting facilities, secure vehicle storage, shipping and receiving centers, offices, and research lab space. (A live database of these tracks is available at www.flpolyami.com.)

Historically, test tracks were the bedrock for validation. However, the increased use of simulation in the context of the V methodology has relegated test tracks to four purposes: characterization, diagnostics, reliability, and environmental effects.

Along with labs, tracks have been used for model characterization such that a far more extensive analysis can be done in simulation. Either in the case of model divergence or physical failure, test tracks have served as the diagnostic function. Finally, reliability and broad environmental effects (cold, rain), which are difficult to model in simulation, constitute the final use-model for tracks.

Table 1 provides a sampling of the important test tracks and their capabilities, such as environmental enclosure, complex weather and street network modeling, virtual simulation, and so on.

In terms of ADS-specific capability, a series of investments have been made in test tracks for physical testing of ADS-enabled vehicles in the U.S., such as Mcity [32, 33]. Mcity is a 32-acre mock city established as a proving ground for the test and validation of CAVs on the University of Michigan North campus in Ann Arbor, Michigan.

In early 2017, the DoT designated 10 locations around the country as proving grounds for self-driving vehicle technology [34]. These are the City of Pittsburgh and the Thomas D. Larson Pennsylvania Transportation Institute, Texas AV Proving Grounds Partnership, U.S. Army Aberdeen Test Center, American Center for Mobility (ACM) at Willow Run, Contra Costa Transportation Authority (CCTA) and GoMentum Station, San Diego Association of Governments, Iowa City Area Development Group, University of Wisconsin-Madison, Central Florida Automated Vehicle Partners, and North Carolina Turnpike Authority. This designation was rescinded in 2018, but these groups continue to exist as entities.

For ADS testing, test tracks can offer a significant utility for various ADS tasks and will continue answering related questions of vehicle manufacturers and regulators in areas such as the following:

- Sensor Calibration: the accurate capturing of data by sensors in various environmental conditions.
- Initial Learning: the correct functioning of a vehicle in a simple, physical environment.
- Characterization: provision of specialized characterization of track testing for simulation environments.
- Diagnostics: the re-creation of a failure in a more controlled environment.
- Reliability: testing reliability through the use of accelerated operation.

Although test tracks provide solutions to specific tasks, they are relatively ineffective in helping solve the key challenge in ADS validation, which is the testing of rare complex scenarios. These scenarios vary in terms of their complexity and are often not understood until ADS-enabled vehicles are exposed to them in the real world. Trying to enumerate them on test tracks is extremely costly and time consuming, and it is likely impossible to generate all these scenarios physically on a test track in any case. Finally, the real issue with ADS testing is the "unknown-unknown" or "known-unknown" problem, and these cannot be handled by static test tracks.

Real-World Driving: Advantages and Disadvantages

With all the issues of test tracks, the common refrain is that this problem will be solved by driving in the real world. At least 12 companies - including BMW, Bosch, Delphi, GM, Ford, Honda, Mercedes-Benz, Nissan, Tesla, Uber, Volkswagen, and Waymo - have been testing their ADS-enabled vehicle fleets on real roads in various locations [35]. ADS testing in the real world is considered the solution complementing the test track validation. However, this is an unsatisfactory solution due to three main reasons: safety, closure velocity, and backward compatibility.

In terms of safety, putting untested vehicles on the road creates safety issues for everyone. The analyses of the ADS failure data show that ADS-enabled vehicles under test are still significantly unsafe compared to the conventional vehicles piloted by humans [36]. Today, the liability framework establishes that the driver is the backup and can, at any time, take over the car. This is the so-called "shadow" driving model. However, this expectation is not very reasonable because it requires the driver to be situationally aware and react instantly upon the detection of trouble. For humans, it is very difficult to be vigilant in situations where most of the time spent is inactive. The backup driver is asked to anticipate a failure not knowing the intention of the AV. Further, rare cases are particularly difficult to test in the real world, given that some of them can be life threatening. Finally, testing manufacturing failures (faults) is not possible in the real world due to safety concerns.

FIGURE 6. **Automotive proving grounds around the world.**

Automotive Proving Grounds

North America

- AET
- American Center for Mobility (ACM)
- Automotive Enviro Testing (AET)
- Pennsylvania Transportation Institute
- Contra Costa Transportation Authority (CCTA) & GoMentum Station
- Cooper Tire and Vehicle Test Center
- North Carolina Turnpike Authority
- Mcity
- International Transportation Innovation Center (ITIC)
- Fowlerville Proving Ground
- Dearborn Development Center
- Eaton ProvingGround
- Fiat-Chrysler Chelsea Proving Ground
- Ford EVE Arizona Proving Ground
- Ford Michigan Proving Ground
- Global Center for Autovie Performance Simulation
- GM Milford Proving Ground
- GM/Army Yuma Proving Ground
- MBtech Auto Testing Propertis LLC
- Michigan Proving Ground
- MGA Research Center
- Nevada Automotive Test Center
- PMG Test and Research Center
- Ride Solutions
- San Diego Association of Governments
- Smithers Winter Proving Ground
- Texas AV Proving Grounds Partnership
- THEA CV Pilot
- Thompson's Subzero Proving Ground
- Transportation Research Center (TRC)
- Toyota Arizona Proving Ground
- University of Wisconsin-Madison
- U.S. Army Aberdeen Test Center
- Virginia Tech Transportation Institute
- Yuma Proving Ground

Europe

- ACTS
- Aldenhoven Testing Center
- Arctic Driving Center
- Asta Zero
- ATP Papenburg
- Aberdeen Test Center
- Brungtingthorpe PG
- BOSCH Juvincourt (France)
- BOSCH Winter Test Center
- Bosch Worldwide
- Colmis Proving Ground
- Icemakers
- IDIADA
- Jules Verne Climatic Wind Tunnel
- Magna Steyr
- MB-technology NA LLC
- MIRA
- MICHELIN Engineering & Services
- Millbrook
- mm-lab GmbH
- Nardo Technical Center
- Navistar
- Nürburgring
- QinetiQ
- Test World
- TRANSPOLIS
- UTAC CERAM
- Winter Testing Finland

Rest of the World

- Australian Automotive Research Center
- Automotive Research & Testing Center, Taiwan
- India National Automotive Testing Track
- Catarc P.G.
- Shanghai GM P.G.
- Southern Hemisphere
- PC Ghang'An P.G.
- WABCO
- You Yangs P.G.
- K-City
- Test Trak

TABLE 1. Test tracks and features.

Test track	Env. enclos.	Urban canyon, weather	Street network	Virtual sim.	Modular facade system	Pick up drop-off	Observ. tower	Pavilion, admin
ACTS [37]		✓						✓
AET [38]		✓	✓					✓
ATC [39]	✓	✓	✓				✓	✓
ADC [40]		✓	✓				✓	✓
AF [41]		✓	✓				✓	✓
APL [42]	✓	✓						✓
ATS [43]		✓						✓
AstaZero [44]		✓	✓				✓	✓
ATP Pap [45]		✓	✓					✓
AARC [46]		✓	✓					✓
ARTC [47]		✓	✓				✓	✓
BoschJuv. [48]			✓	✓				✓
Bosch WTC [49]		✓	✓	✓				✓
BPG [50]			✓					✓
Changan [51]			✓			✓		✓
CPG [52]	✓	✓	✓					✓
CTVTC [53]		✓	✓					✓
FPG [54]			✓					✓
GTF [55]		✓						✓
GCAPS [56]								✓
GS Eng. [57]		✓						✓
Icemakers [58]		✓				✓	✓	✓
IDIADA [59]		✓					✓	✓
JVCWT [60]	✓							✓
MB-tech NA [61]	✓	✓		✓				✓
Mcity [32]	✓		✓	✓	✓	✓		✓
MIRA [62]		✓	✓	✓			✓	✓
MICHELIN Eng. [63]		✓						✓
Michigan [64]							✓	✓
Michigan TRP [65]								✓
Millbrook [66]							✓	✓
mm-lab [67]							✓	✓
NATRIP [68]							✓	✓
NATRAX [69]							✓	✓
NardoTC [70]	✓							
Navistar [71]							✓	✓
Nrburgring [72]			✓		✓		✓	✓
Nevada ATC [73]	✓						✓	✓
PMG Tech [74]					✓		✓	✓
QinetiQ [75]	✓							✓
Ride Sol. [76]	✓	✓						✓
SWTC [77]								
SHPG [78]								✓
TestWorld [79]	✓							✓
Thompson SZ [80]	✓							✓
TRANS POLIS [81]			✓	✓		✓		✓
TRC [82]		✓	✓	✓				✓
UTAC CERAM [83]	✓	✓	✓	✓		✓		✓
Uvalde [84]				✓				✓
Wabco [85]						✓	✓	✓
Waymo	✓	✓	✓	✓	✓	✓	✓	✓
WT Finland [86]	✓							
Yangs [87]								✓
Yuma [88]				✓				

FIGURE 7.

FIGURE 8.

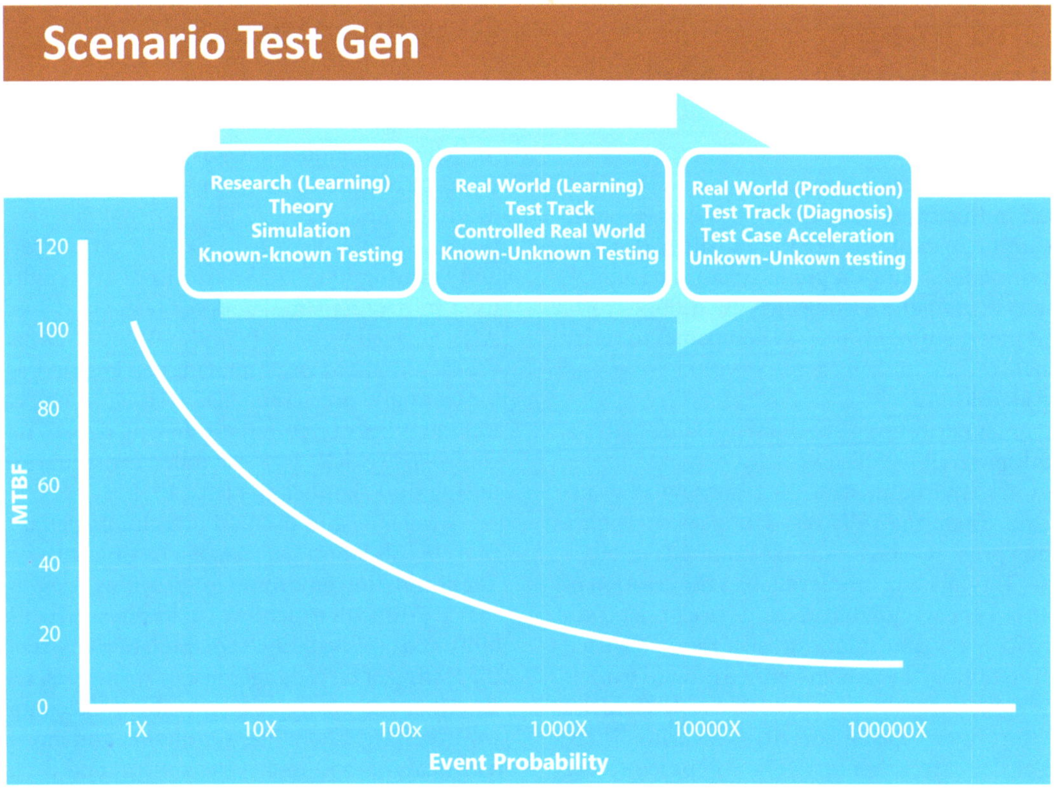

Closure velocity is also a significant disadvantage of real-world driving tests. That is, real-world testing is painfully slow and expensive in terms of exposing an AV to interesting scenarios. Thus, the closure velocity to validation progresses very slowly. To achieve a confidence level of ADS safety, one would need to handle six-sigma events; however, it would take too long to reach these events in the real world [8]. What are examples of six-sigma events? A cat crossing the road in front of your car. If one looked at all the miles driven by cars, this is a "rare" event. However, with over 200 million cars on the road today, it is likely this happens every day somewhere. The number of manufacturers involved and number of ADS-enabled vehicles on the road are increasing; yet, the total number of ADS-enabled vehicles is still significantly low compared to the total number of conventional vehicles currently on the road.

In any testing environment, backward compatibility is critical. That is, the next version of the car should prove safer than the previous version. Due to the nature of the current AI learning engines and real-world driving, it is not clear whether a new generation of AI/ML engine is actually better in all the important dimensions in comparison to an older version. For example, imagine an AV has learned how to handle the very difficult situation of discerning between a leaf and a stone on the road. How does one know that this shift in learning has not adversely affected some other previously working situation? In most software and hardware systems, these situations are solved by a regression test and signoff process before release. As bugs are found and fixed, they are added to the regression suite. What is the regression signoff for AV systems? Today, a majority of the reported disengagements are the results of AI decisions [89]; yet, no objective measurement method is used to quantify the improvements nor are there any guarantees of backward compatibility.

Video and Image Datasets: Advantages and Disadvantages

Simulation and emulation have been added as a tool for ADS validation, and in this case, video and image datasets have been used as an important tool. The tests with video and image datasets include labeling the ground truth and running the perception systems of the ADS against this ground truth. This method is most commonly used by companies to verify the changes that are done in their computer vision, radar, and deep learning algorithms.

The companies performing ground truth labeling typically have multiple ADS-enabled vehicles driving in real-world or test-track environments while recording all available sensor data. Then, the collected data goes through a manual labeling process, which can take hundreds or thousands of hours. This labeling process enables the creation of a ground truth, where every detail of the device under test's drive is known for every time stamp. Because the data is used for safety-critical testing, the human-verified ground truth is used. After the ground truth is created, ADS-enabled vehicles are tested against query-specific driving scenarios. Then, the behavior of the perception system is compared to the expected results. The ground truth also allows the system's characteristics to be recorded according to their behaviors in different conditions, such as weather and road conditions.

Several highly detailed video benchmarks exist to evaluate ADS computer vision approaches [90]. One of these benchmarks, Cityscapes Dataset [91], contains stereo video sequences recorded in 50 different cities with 5,000 pixel-level and 20,000 weakly annotated frames. Another benchmark suite example, the "KITTI Vision Benchmark Dataset" [92], includes sparse pixel-level annotations of video streams created by using light detection and ranging (LiDAR) laser scanners.

The video datasets provide problem-specific examples and enable the evaluation and comparison of computer vision methods. However, they are ineffective for ADS testing for various reasons. For instance, most of these datasets require extensive annotation effort for camera and other sensor inputs. Work is being performed to automate this process, but the annotation still requires significant time and effort. Additionally, the video and image datasets have challenges similar to test tracks in terms of the inefficiency in the creation of complex scenarios and corner cases. The recorded data are generally not modified in ADS testing and used as they are, with the exception of a few augmented training scenarios [93]. The test scenarios are limited by the collected data, and it is very difficult to create and test dangerous or rare cases. Therefore, video dataset-based testing has limited usability for system-level scenario testing. Similar to test tracks, static datasets cannot handle the "unknown-unknown" or "known-unknown" problem.

Modeling and Simulation: Advantages and Disadvantages

Given the constraints of test tracks, real-world driving, and video datasets, there has been momentum in the area of modeling and simulation. For instance, Waymo [94] complements its real-world driving tests by simulating three million driving miles every day [95]. Virtual datasets of synthetic images have been created for use in various forms of testing and training [96, 97]. These simulated environments are used in driving simulators for tests on drivers, ADS, and ADS-enabled vehicles [98, 99, 100, 101]. Other approaches use the simulated environments of computer games as a basis for testing to accelerate ADS testing, in one case using 1.3 million miles of naturalistic driving data to model human-controlled vehicles for a car-following scenario to reduce the testing time significantly [102, 103].

Modeling and simulation technologies are mostly used in ADS validation for scenario testing with synthetic data. The simulation environment provides two critical features that are difficult to achieve in the physical world: controllability and observability. The simulation environment enables the testing of rare, dangerous, or corner cases, which would be either too costly or impossible to accomplish in real-world testing. Therefore, simulation and modeling provide a critical method to test the control and decision-making systems of ADS-enabled vehicles, in particular. Most importantly, simulation enables the V paradigm, whereby components such sensor systems (LiDAR, vision, etc.) and object-recognition systems can be tested in a simulation and then verified with hardware-in-the-loop techniques.

Table 2 offers insight into simulation platforms around the world with their origin (academia or industry), organization (company, university, research institute, etc.), and code availability (open source or proprietary). The table also shows the focus and two other main features for each platform. The majority of these platforms can be grouped as

TABLE 2. Simulation platforms for AV verification.

Platform	Origin	Organization	Source	Focus	Features
3xD Simulator [98]	Academia	Uni. of Warwick	Prop.	Driving simulation	360° View, infotainment HiL
Aimsun Simulator [104]	Industry	AimSun	Prop.	Traffic simulation	Network top., vehicle dynamics
AirSim [105]	Industry	Microsoft	Open	Driving simulation	Unreal plugin, PX4 controller
Apollo Simulation [106]	Industry	Apollo, Baidu	Open	Driving simulation	Scen. gen., grading system
ASIM Driving Simulator [107]	Academia	Virginia Tech.	Prop.	Driving simulation	V2X, brain monitoring
AutonoViSim [108]	Academia	UNC,UCF	Prop.	Driving simulation	Strategy testing, modular struct.
AutoSim AS 1200-6 [109]	Industry	MTI (Poland)	Prop.	Driving simulation	200° view, 6 DoF
AV Simulation [110]	Industry	Cesium	Prop.	Visualization	3D mapping, sensor cov. analysis
AV Simulator [111]	Industry	monoDrive	Prop.	Driving simulation	Scen. gen., vehicle dynamics
Blender [112]	Industry	Blender	Open	Visualization	Dynamic 3D models, animation
BMW Driving Simulator [113]	Industry	BMW	Prop.	Driving simulation	360° View, 7 Chan. Projection
CarCraft [94]	Industry	Waymo	Prop.	Driving simulation	Real-life scen. replay, scen. gen.
CARLA [114]	Academia	UAB,Intel, Toyota	Open	Driving simulation	Scen. Gen., Sensor Suite
Cognata Sim. Engine [115]	Industry	Cognata	Prop.	Driving simulation	Whole-city sim., traffic models
CORSIM [116]	Academia	UF,McTrans	Prop.	Traffic simulation	Network top., vehicle dynamics
Drive Constellation [117]	Industry	NVIDIA	Prop.	Driving simulation	Sensor suite, scen. gen. HiL
DRIVE-IN [118]	Academia	U.Porto, CMU Port.	Prop.	Driving simulation	VANET integration, V2V
Driving Simulator [119]	Academia	Uni. of Leeds	Prop.	Driving simulation	human reaction analysis, scen. gen.
Driving Simulation [120]	Industry	rFpro	Prop.	Driving simulation	Road models, scen. gen.
dSpace [121]	Industry	dSpace	Prop.	Driving simulation	MIL, SIL, HIL toolchain
Duckietown [122]	Academia	MIT	Open	Robotics simulation	Gazebo integration, robot control
Dynamic Driving Sim. [100]	Academia	DLR (Germany)	Prop.	Driving simulation	270° view, 9 chan. projection
Gazebo [123]	Industry	Gazebo	Open	Robotics simulation	3D models, sensor models
iCAVE2 [124]	Academia	Uni. of Buffalo	Prop.	Driving simulation	Traffic and network simulation
MATLAB (and Simulink) [125]	Industry	MathWorks	Prop.	Driving simulation	Scen. gen., ground-truth labeling
MATSim [126]	Academia	ETH Zurich	Open	Traffic simulation	Network top., vehicle dynamics
Metamoto Simulator [127]	Industry	Metamoto	Prop.	Driving simulation	Parametrization, Multi-Sensor Sim.
Miss Daisy [128]	Academia	MIT	Prop.	Driving simulation	150° view, modular interface
MORSE [129]	Academia	LAAS CNRS	Open	Robotics simulation	Portable, 3D models
NADS [130]	Academia	Uni. of Iowa	Prop.	Driving simulation	Large and small-scale simulators
OpenDRIVE [131]	Industry	VIRES	Open	Driving simulation	Road network planning, scen. gen.
OpenEnergySim [132]	Academia	NII Japan	Prop.	Driving simulation	Online, multiple actor roles for users
OSU Driving Simulator [133]	Academia	OSU, Honda	Prop.	Driving simulation	260° view, 6 DoF
Paramics [134]	Industry	SYSTRA	Prop.	Traffic simulation	Network top., vehicle dynamics
PTV Vissim [135]	Industry	PTV Group	Prop.	Traffic simulation	Network top., vehicle dynamics
RightWorld [136]	Industry	RightHook	Prop.	Driving simulation	Scen. Gen., Vehicle Dynamics
RoadView [137]	Academia	Jiaotong Uni.	Prop.	Driving simulation	Image sequence, GIS data integration
Scaner [138]	Industry	Oktal, Renault	Prop.	Driving simulation	Road network, visual-audio platform
Self-Driving Car Sim. [139]	Industry	Udacity	Open	Driving simulation	Modular, physiological measurements
SHERPA [140]	Academia	Univ-Valenci ennes, Peugeot	Prop.	Driving simulation	240° view, 6 DoF
SimDriver [141]	Industry	Realtime Tech.	Prop.	Driving simulation	Human reaction analysis, scen. gen.
simRING [142]	Academia	Uni. of Buffalo	Prop.	Driving simulation	Motion platform, 4 screen projection
Simulation at FCAV [143]	Academia	UMich	Prop.	Driving simulation	Replacing annotations, scen. gen.
Stage [144]	Academia	SFU	Open	Robotics simulation	Environment and sensor models

TABLE 2. (Continued) Simulation platforms for AV verification.

Platform	Origin	Organization	Source	Focus	Features
Stanford Driving Sim. [145]	Academia	Stanford, Toyota	Prop.	Driving simulation	Brain EEG, programmable dash
STISIM Drive [146]	Industry	Systems Technology	Prop.	Driving simulation	Scen. gen., plug-in modules
Stuttgart Simulator [99]	Academia	Stuttgart University	Prop.	Driving simulation	360° view, 12 projectors
SUMO [147]	Academia	DLR (Germany)	Open	Traffic simulation	Portable, network top.
SynCity [148]	Industry	Cvedia	Prop.	Driving simulation	Sensor suite, scen. gen.
Tass [149]	Industry	Siemens	Prop.	Driving simulation	Scen. gen., sensor suite
Toyota Driving Simulator [150]	Industry	Toyota	Prop.	Driving simulation	360° view, 12 DoF
TROCS [151]	Academia	CMU	Prop.	Driving simulation	Offline replay, operator interface
Uber Visualization [152]	Industry	Uber	Prop.	Visualization	Web-based, data fusion
USARSim [153]	Academia	UC Merced	Open	Robotics simulation	UnReal based, scen. gen.
V-Rep [154]	Industry	Coppelia Robotics	Prop.	Robotics simulation	Portable, sensor models
VIRTTEX [155]	Industry	Ford	Prop.	Driving simulation	360° view, 24 feet dome
Virtual Driv. Platform [156]	Industry	Daimler	Prop.	Driving simulation	Scen. gen., driving platform
Virtual Test Drive [157]	Industry	VIRES	Prop.	Driving simulation	Network top., scen. gen.
VTI Simulator [158]	Academia	VTI	Prop.	Driving simulation	360° view, 6 DoF
WISC Driving Simulator [159]	Academia	WISC	Prop.	Driving simulation	240° view, 7 feet tall screen

"driving simulators." While some are large research facilities housing platforms with several degrees of freedom (DoF), others are small-scale simulators with scenario generation (scen. gen.) capabilities. Microscopic traffic simulators are also widely used to design and test solutions in transportation network topologies.

Some initiatives, intended to facilitate interoperability, define integration mechanisms for multiple methodologies. Ultimately, the goal of these approaches is the standardization of ADS testing procedures. One of these approaches, the Intelligence Testing Framework [160], defines a semantic diagram that explains the relationship between testing scenarios, tasks, and vehicle functions. This diagram is then traversed to generate tests, which aim to find a link between the scenario and the required tasks.

The Project for the Establishment of Generally Accepted quality criteria, tools, and methods, as well as Scenarios and Situations for the release of highly automated driving functions (PEGASUS) [161, 162] is a large joint effort to define a standardized validation procedure with 17 main and 4 affiliated partners from academia and industry. PEGASUS has four main subprojects. The "Scenario Analysis and Quality Measures" subproject aims to define the human and technical capacity requirements and related criteria for ADS-enabled vehicles. The "Implementation Processes" subproject investigates the required testing tools and methods. The "Testing" subproject focuses on coverage of tests, their measures, and the role of different levels, such as simulation, proving grounds, and labs. The "Result Reflection and Embedding" subproject concentrates on the sustainability and transferability of the proposed concepts.

PEGASUS and Intelligent Testing Framework are examples of modern ADS testing solutions designed to start tackling the outlined key challenges.

In terms of disadvantages, nearly all the simulation work done today relies on manual test generation or very limited situational analysis with automation (e.g., Google pedestrian tester). The manual test generation in simulation suffers from three significant issues: scalability, test generation bias, and criteria for success. As a result of the required time, cost, and scenario knowledge, manual test generation limits the number and variety of complex scenarios that are created.

Because tests are built by humans, the tests often have bias relative to the expected operation of the world, and this phenomenon is especially prevalent when designers design the tests. In the case of ADS-enabled vehicles and the most interesting test cases, the criteria for success is the expected action of humans, which must be defined. In summary, simulation offers a very useful platform for addressing the key issues for ADS validation, but issues such as "criteria for success," automated test generation, and methodology for constant improvement remain vexing issues.

Recommendations

There is a place for conventional test tracks, real-world testing, and video/image datasets. However, to handle the validation tasks, it is likely each will have to move into a more integrated flow relative to the overall validation task. What does this mean?

1. Real-World Testing: This form of testing will continue to have value to show viability. However, this testing will be most valuable when coupled within an overall validation framework where the real-world driving test is guided by desired scenario conditions generated by other means.

2. Video and Image Datasets: These are likely to be useful in the context of a regression testing environment which is augmented within the context of an overall validation flow.
3. Test Tracks: The most successful and useful test tracks will focus on providing physical assets that can quickly and efficiently build scenarios. These modern test tracks will look much more like a movie set as opposed to a traditional racetrack. In addition, the level of testing capacity available for EMI issues as well as local information technology (IT) resources will be critical. The costs of these tracks will be significant, so a network of cooperating tracks may well be required.

Overall, given the immense complexity of the AV validation task, nearly all methods will used.

Modeling and Simulation are very likely to be the bedrock of a future AV validation structure. However, to be helpful, industry cooperation is needed on many fronts.

1. Scenario Abstraction: An agreement on the basic format and characteristics of a scenario abstraction would enable the development of third-party tools and the sharing of test sets.
2. Test Generation: Given the need to compress validation learning, it is vitally important to generate the "right" tests. Often known as "edge" conditions, significant research is required to generate these tests as automatically as possible.
3. Regression: With every release of an updated AV system, the need exists to develop a testing methodology that assures that the system has not degraded from a safety point of view. Thus, an efficient regression structure is necessary.

Hardware Validation and Natural Language Processing

The current methods used for ADS validation fall short in addressing the current challenges. Thus, a paradigm shift is needed, and several research directions focus on the solution to these issues in other similar fields.

One mapping pursued by several players, such as Florida Polytechnic University's Advanced Mobility Institute, is to use the techniques from hardware validation and Natural Language Processing (NLP). In this section, we explore these fields with the idea of mapping their solution matrix to that of ADS validation and in turn proposing an initial framework for ADS validation.

Similar Challenges from a Different Field: Hardware Validation and Natural Language Processing

The validation of complex digital hardware such as a modern central processing unit (CPU) chip has challenges very similar to the challenges in ADS validation.

1. Complex Underlying Behavior: The fundamental underpinnings of a chip are semiconductor devices, which have complex physics behavior. Just modeling a single transistor and simulating its detailed behavior can take considerable time and resources. The actual chip is built using millions of these semiconductor devices interconnected in complex ways.
2. Learning Gap: The chips themselves run at gigahertz (109) rates while the best-case simulation performance is in the thousands of hertz, and even hardware-assisted methods reach only into the megahertz. The implication of this disparity is that if one day's worth of activity is simulated, this only amounts to 86 seconds worth of chip activity [163].
3. High Cost of Failure: The cost of failure is very high because of the manufacturing costs of semiconductors. A simple hardware update may well cost millions of dollars.

All these problems also exist in ADS validation. The hardware validation world has come up with the following three key methods to solve these problems such that most modern chips exit the semiconductor fabrication plant working correctly:

- Abstraction and separation of concerns,
- Constrained pseudo-random test generation, and
- Assertions and coverage analysis.

Within the hardware design and validation process, there is a strong use of abstraction to separate concerns and decompose the process of validation.

The fundamental abstractions in the world of hardware design include layout, transistors, register-transfer level (RTL), and high-level protocols. Each level is characterized with a self-consistent formal definition with execution semantics and chip validations are performed in two distinctive techniques:

- Equivalence: Two levels of abstractions are shown to be equivalent to each other.
- Top Level validation: The top level is verified against the intended specification.

The combination of equivalence and top-level validation is a form of proof by induction. If the top level is correct and the equivalence with the downstream abstractions can be shown, the final product will work as intended. In hardware design, the equivalence problem has been shown to be much easier compared to the top-level validation. In many cases, formal methods or near-exhaustive component level simulation has been effective for equivalence. However, the top-level validation has been a much more difficult challenge. Thus, in addition to conventional real-world simulation of program streams, hardware validation has made heavy use of the following techniques:

- Constrained pseudo-random test generation,
- Coverage-driven testing, and
- Functional assertions.

The result of using these techniques has been a remarkable success in first time functioning silicon. When there is a coverage hole, this methodology lends itself to incremental update.

In hardware validation, the "spec" is a human-defined behavior, which is typically a programming model. This can be defined to some degree in the world of ADS-enabled vehicles. As a simple example, the "spec" can be "keep a safe distance to all other objects." However, the actual "spec" for an ADS-enabled vehicle is to act like a good human driver. In this way, ADS validation introduces an additional challenge beyond traditional hardware validation. This largely comes from the area of the right "assertions," which need to be included into the system. That is, "What is the criteria for success?"

In order to address the "assertions" aspect of the validation problem, the techniques from NLP, which aims to understand unwritten rules of human behavior, can be used. The history of NLP has been to glean function from machine learning over a population of humans [164]. This technique can also be used to glean out assertions for the safety system. That is, correctness is determined by a machine-learned method for determining the correct behavior based on observations of humans. As an interesting consequence, this methodology also enables an implicit deduction of the environment. As an example, one can deduce there is a stop sign based on the behavior of humans to a location as opposed to the explicit recognition of a stop sign.

Mapping Hardware Validation Solutions to Autonomous Vehicles

The main concepts of hardware validation can be instrumental for testing and validation of ADS-enabled vehicles. The abstraction stack for ADS-enabled vehicles consists of three layers: Raw Sensor Data (bits/bytes), Object Recognition, and Planning and Decision-Making. Adopting the "separation of concerns" principle, the ADS validation for these layers can be separated.

- Sensor Reliability/Calibration: This is a critical part of the validation problem and it can be solved at the component level.
- Object Recognition: The problem of object recognition requires a larger number of inputs compared to sensor calibration. However, it can also be solved at a high level of granularity.
- Decision-Making: The validation of decision-making is a complex problem, which requires a formalization of the scenario level abstraction.

It is important to note that we focus on the validation of the planning and decision-making aspect of the ADS in this paper. The principles and rich machinery of hardware validation can be utilized in this space. In order to do so, a new level of abstraction must be defined.

This definition of scenario abstraction enables a powerful decomposition flow, as shown in Figure 9. In this flow, constrained random scenarios can be generated automatically. The range of these can help address unknown-unknown and known-unknown situations (as they do in hardware validation). The signature of the scenario can be used for building a coverage matrix that allows for accumulative knowledge on safety and connects with machine learning for determining criteria for success and backward compatibility. The scenario generation can serve as the ground truth for the independent process for sensor/object recognition testing. Real-world crashes can be abstracted into this environment and can generate test cases for test tracks for diagnostic purposes. Finally, the system provides a technical language for industry and regulators to clearly communicate failure information.

Recommendations

Hardware verification offers many lessons that can aid the process in AV validation. Critical aspects include the concepts of abstraction, separation of concerns, and machinery for automated test generation. All these concepts have a great deal of utility in the world of AV validation.

Progressive Structures: Evolving Requirements to Ensure Future Safety

It is very likely new paradigms of modeling will be developed for specification and validation to overcome challenges

FIGURE 9.

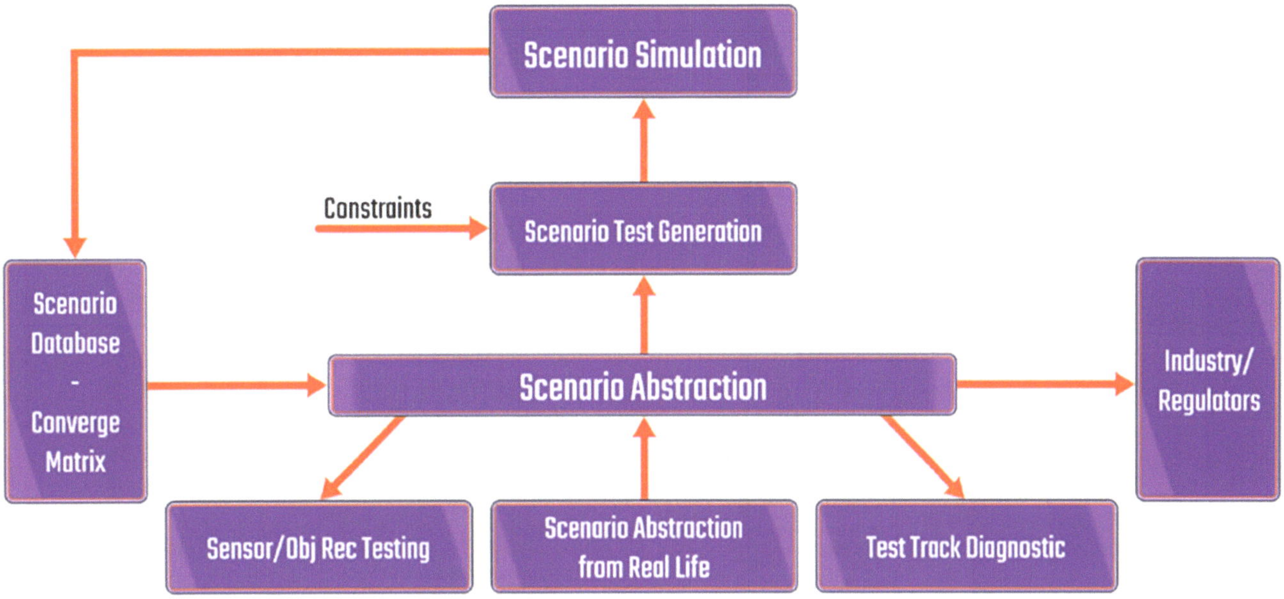

and work through unsettled domains. All players in the ADS ecosystem will face and need to react to new challenges.

Original Equipment Manufacturers (OEMs)

For OEMs, the shift from a conventional vehicle to an ADS paradigm marks a massive increase in complexity. OEMs will have to consider the following points.

- Does it make sense to standardize on an internal architecture that separates major functions in an ADS ecosystem and exposes the internal data as an observation point?
- Does it make sense to standardize test data (especially edge case scenario data) such that the broader community can increase safety?
- Will ADS-enabled vehicles have to be released micro-regionally to account for the driving differences between regions or even in the same country?
- Will it be possible to build sufficient testing capacity internally or is the task so large that there must be access to a network of testing resources shared by the community?
- Do the economics of ADS-enabled vehicles work at scale and how does one get to this scale?
- Given the rising costs of test and validation, does it make sense to build shared multiuse facilities?

Validation Services Providers

Validation services providers can be generally split into the providers of software solutions and physical track resources. For the software providers, the critical issues will be:

- How do you trade off various levels of fidelity/detail versus simulation performance?
- How do you aid in the generation of interesting test cases independent of level of fidelity and how is this measured?
- How does the software solution scale in terms of running millions or perhaps billions of test cases?

For physical track providers, a static track with little regard to control ability and observability is not likely to be very useful. The critical issues for physical track providers will be:

- How does one build flexibility (like a movie set design) such that scenarios from simulation or real-life accidents can be built and evaluated quickly?
- How does one build an infrastructure to manage the electronic aspects of an ADS-enabled vehicle (noise generation from any direction, visibility from any direction, dynamic movement of all electronics pieces)?
- What is the data center and network model to move these capabilities back and forth from design to test?

Standardization Bodies and Regulators

For standardization bodies such as SAE International, the Institute of Electrical and Electronics Engineers (IEEE), and others, the challenge is to wait for the technology to settle and then standardize to create and support an active ecosystem. Level 3 through Level 5 are not yet sufficiently mature to start a standardization task; however, Level 2 is ripe for standardization on many aspects of the solution, including testing methodology.

Regulators must similarly lag the development of the technology in order to understand the problem sufficiently to regulate it. However, some high-level thrusts seem to make sense.

- Regulators need to consider physical (traditionally NHTSA), electronic [traditionally Federal Communications Commission (FCC)], and weather [traditionally National Oceanic and Atmospheric Administration (NOAA)] regulation and interaction. Radio frequency (RF) technology is being contemplated, potentially for the first time in a safety-critical application, for ADS-enabled vehicles.
- Regulators need to drive the technical community for the appropriate technical frameworks and languages that can be the basis of regulation.
- Regulators will likely need to harmonize with local driving behaviors in a much deeper way. Juneau in Alaska, Manhattan, and Los Angeles have very different ADS-enabled vehicle requirements, for example.

Recommendations

The road to ADS adoption is paved with highly significant challenges, for all stakeholders, which likely cannot be solved with traditional information technology (IT) solutions. Safety considerations are paramount. The best way forward involves simplification, cooperation, and standardization.

1. Simplification: Focusing on simplified use models as a community would greatly accelerate progress. As an example, public transportation low-speed, fixed-route solutions provide value and can be a great step toward developing an overall ecosystem. Lessons learned can be methodically applied and improved for more complex situations.
2. Cooperation: Engagement and active alignment around safety issues will be critical for everyone's success. It is in the industry's best interest to cooperate, collaborate, and jointly support safety-related efforts and overcome safety challenges and issues. A propriety advantage that could have saved lives in other vendors' solutions will likely not be viable in the long run.
3. Rational scientific validation structure: Without a standardized rational validation structure, which has the characteristics of constant learning, it will be very difficult to instill confidence and gain public trust.

Summary/Conclusions

ADS technology will have a significant impact on multiple industries as soon as the next decade. These industries include not only the traditional markets, such as passenger vehicles, but also nontraditional markets, such as logistics, agriculture, planned communities, and public transportation. Consequently, nearly every sector of economy will be affected directly or indirectly from this form of autonomous transformation. Yet, the validation of ADS-enabled vehicles for full autonomy remains an open challenge critical to complete the transformation.

Next Steps for Autonomous Vehicle Test and Validation

This publication should be considered only as a first step toward clarifying the issues around autonomous vehicle testing and validation. The intention behind this and other SAE EDGE™ Research Reports is to start a dialogue among interested parties on important industry-wide topics that require further attention. The expectation is that these explorations of unsettled areas of technology will lead to the formation of working groups and, ultimately, committees that can address and resolve the issues they raise, producing a framework for developing a common vocabulary of definitions, best practices, protocols, and standards needed to support continued progress toward safer and more innovative automotive products.

The experts' meeting that gave rise to this publication demonstrated a great willingness on the part of the industry to define the terminology, procedures, and eventually the standards needed to enable autonomous vehicle testing and validation technology to move ahead as quickly and efficiently as possible. SAE International has demonstrated its lead in this and closely related areas by the enabling the communication of the critical issues. The next stage will be the formation of standards and consortia.

This SAE EDGE™ Research Report on autonomous vehicle testing and validation identifies the following key topics for further pursuit, both through continued informal discussions among industry practitioners and through more formal working groups:

- Replacing the driver, compressing human learning and understanding unwritten rules of the road

- Regulations, which inform and affect ADS test and validation approaches
- Finessing existing validation approaches to fit new ADS paradigm
- Hardware validation and natural language processing
- Progressive structures: evolving requirements to ensure future safety

Recommendations

The overall recommendations of this SAE EDGE™ report can be summarized as follows:

1. AV validation requires a framework that accounts for the additional tasks of perception and decision-making, which are critical to automated and autonomous systems.
2. Current validation methods, ranging from test tracks to simulation, will need to be significantly upgraded in order to handle the challenge.
3. Simplification, cooperation, and standardization are likely to be the key methods to enable progress.
4. The use of RF technology (V2V, V2I) for safety-critical issues must be studied very closely for interference issues from natural and human-made sources.
5. Cross-learning from fields ranging from natural language processing, hardware verification, aircraft validation, and more will likely be part of the optimal solution.
6. Significant challenges beyond those technical in nature exist and must be overcome for industry to achieve the full economic potential of ADS solutions. Stakeholders are faced with a chicken-and-egg scenario, around whether to first invest in semiconductor integration or civil infrastructure.

SAE EDGE™ Research Reports

SAE EDGE™ Research Reports, like the present report on "Unsettled Technology Areas in Autonomous Vehicle Test and Validation," are intended to push further out into still unsettled areas of technology of interest to the mobility industry. SAE launches these reports before attempting to form a joint working group, let alone a CRP or a Standards Committee.

SAE EDGE™ Research Reports are intended to be quick, concise overviews of major unsettled areas where vital new technologies are emerging. By definition, an unsettled area is characterized more by confusion and controversy than established order. Early practitioners must confront an absence of agreement. Their challenge is often not to seize the high ground but to find common ground. These scouting reports from the frontiers of investigation are intended merely to begin the process of sorting through critical issues, contributing to a better understanding of key problems, and providing helpful suggestions about possible next steps and avenues of investigation.

SAE EDGE™ Research Reports, therefore, are fundamentally distinct from the more formal working-group approach and far removed from the more mature CRP and standard's development process. They begin by convening an informal meeting of subject matter experts working in an unsettled area of technology from industry, government, and the research community. A moderated but collegial discussion ensues regarding their shared interest. This conversation attempts to identify critical issues hindering a clear understanding of important topics, preventing a common agreement on potential solutions, or slowing progress in key areas of research.

An SAE EDGE™ Research Report is then drafted and circulated for comment to meeting attendees and other interested parties. After review, the final result of this process is a report like the current examination of unsettled technology areas in autonomous vehicle test and validation.

Abbreviations/Definitions

AI - Artificial Intelligence. Computer systems that can exercise judgment and make decisions; e.g., recognize objects from visual inputs, understand spoken or written words and translate between languages, make optimal choices between alternative moves relative to achieving a goal (e.g., playing chess, navigating), etc.).

ACM - American Center for Mobility

ADS - Automated Driving System

CAV - Connected Automated Vehicle

CCTA - Contra Costa Transportation Authority

CPU - Central Processing Unit. Used interchangeably with processor or microprocessor, though modern microprocessors incorporate far more functionality than just the central processing unit responsible for executing program instructions.

CV - Connected Vehicle

DoF - Degrees of Freedom

DoT - Department of Transportation. The US federal agency that oversees interstate travel.

EMI - Electromagnetic Interference

FCC - Federal Communications Commission

GPS - Global Positioning System. Use of satellite signals to determine the surface position of GPS-equipped objects on the earth, with an average accuracy under 4 meters of horizontal displacement.

IEEE - Institute of Electrical and Electronics Engineers

IIHS - Insurance Institute for Highway Safety

ITS - Intelligence Transportation Systems

LiDAR - Light Detection And Ranging.

NHTSA - National Highway Traffic Safety Administration. The US federal agency charged with writing and enforcing federal motor vehicle safety standards.

NLP - Natural Language Processing

NOAA - National Oceanic and Atmospheric Administration

PEGASUS - Project for the Establishment of Generally Accepted quality criteria, tools and methods as well as Scenarios and Situations for the release of highly automated driving functions (from the German *Projekt zur Etablierung von generell akzeptierten Gütekriterien, Werkzeugen und Methoden sowie Szenarien und Situationen zur Freigabe hochautomatisierter Fahrfunktionen*). A joint project promoted by the German Federal Ministry for Economic Affairs and Energy (BMWi) to deliver standards for automated driving.

OEM - Original Equipment Manufacturer

RF - Radio Frequency

RTL - Register-Transfer Level

V2I - Vehicle-to-Infrastructure

V2N - Vehicle-to-Network

V2V - Vehicle-to-Vehicle

V2X - Vehicle-to-Everything

Works Cited

1. Waldrop, M.M., "No Drivers Required," *Nature* 518 (2015): 20–20.
2. Shoup, D.C. et al., *The High Cost of Free Parking*, Vol. 206 (Chicago, IL: Planners Press, 2005).
3. Lovins, A., *Reinventing Fire: Bold Business Solutions for the New Energy Era* (White River Junction, VT: Chelsea Green Publishing, 2013).
4. National Safety Council (NSC) Statistics Department, "NSC Motor Vehicle Fatality Estimates," 2017, accessed June 1, 2019, https://www.nsc.org/portals/0/documents/newsdocuments/2018/december_2017.pdf.
5. Treat, J.R., Tumbas, N., McDonald, S., Shinar, D., and Hume, R., "Tri-Level Study of the Causes of Traffic Accidents. Volume I: Casual Factor Tabulations and Assessments," Technical Report, 1977.
6. Uber, "Driving with Uber, How Much Do Drivers with Uber Make?," 2017, accessed June 1, 2019, https://www.uber.com/info/how-much-do-drivers-with-uber-make/.
7. Fagnant, D.J. and Kockelman, K., "Preparing a Nation for Autonomous Vehicles: Opportunities, Barriers and Policy Recommendations," *Transportation Research Part A: Policy and Practice* 77 (2015): 167–181.
8. Kalra, N. and Paddock, S.M., "Driving to Safety: How Many Miles of Driving Would It Take to Demonstrate Autonomous Vehicle Reliability?," *Transportation Research Part A: Policy and Practice* 94 (2016): 182–193.
9. Kröger, L., Kuhnimhof, T., and Trommer, S., "Does Context Matter? A Comparative Study Modelling Autonomous Vehicle Impact on Travel Behaviour for Germany and the USA," *Transportation Research Part A: Policy and Practice* 122 (2018): 146–161.
10. Talebpour, H. and Mahmassani, S., "Influence of Connected and Autonomous Vehicles on Traffic Flow Stability and Throughput," *Transportation Research Part C: Emerging Technologies* 71 (2016): 143–163.
11. Mersky, C. and Samaras, C., "Fuel Economy Testing of Autonomous Vehicles," *Transportation Research Part C: Emerging Technologies* 65 (2016): 31–48.
12. Sivak, M., "A Survey of Public Opinion about Autonomous and Self-Driving Vehicles in the U.S., the U.K., and Australia," 2019, [online] accessed June 6, 2019, https://deepblue.lib.umich.edu/handle/2027.42/108384.
13. Fraedrich, E., Heinrichs, D., Bahamonde Birke, F.J., and Cyganski, R., "Autonomous Driving, the Built Environment and Policy Implications," *Transportation Research Part A: Policy and Practice* 122 (2018): 167–172.
14. SAE On-Road Automated Vehicle Standards Committee, "Taxonomy and Definitions for Terms Related to On-Road Motor Vehicle Automated Driving Systems," SAE Standard J3016 (2014): 01–16.
15. Dey, K.C., Rayamajhi, A., Chowdhury, M., Bhavsar, P., and Martin, J., "Vehicle-to-Vehicle (V2V) and Vehicle-to-Infrastructure (V2I) Communication in a Heterogeneous Wireless Network–Performance Evaluation," *Transportation Research Part C: Emerging Technologies* 68 (2016): 168–184.
16. HomChaudhuri, B., Lin, R., and Pisu, P., "Hierarchical Control Strategies for Energy Management of Connected Hybrid Electric Vehicles in Urban Roads," *Transportation Research Part C: Emerging Technologies* 62 (2016): 70–86.
17. Litman, T., "Autonomous Vehicle Implementation Predictions," Victoria Transport Policy Institute, 2017.
18. California Department of Motor Vehicles, "Report of Traffic Collision Involving an Autonomous Vehicle (OL 316), accessed June 1, 2019, www.

dmv.ca.gov/portal/dmv/detail/vr/autonomous/autonomousveh_ol316.

19. IEEE Spectrum: Technology, Engineering, and Science News, "Fatal Tesla Self-Driving Car Crash Reminds Us That Robots Aren't Perfect," 2019, [online] accessed June 6, 2019, https://spectrum.ieee.org/cars-that-think/transportation/self-driving/fatal-tesla-autopilot-crash-reminds-us-that-robots-arent-perfect.
20. IIHS, "Tests Uncover Issues for Advanced Features," accessed June 1, 2019, www.iihs.org/iihs/news/desktopnews/evaluating-autonomy-iihs-examines-driver-assistance-features-in-road-track-tests.
21. Hariga, F.-R., Cuza, A.I. et al., "A Problem of Context in Defining Psychopathy: A Flaw of Reason or a Flaw of Will?, *Analysis* 2 (1997): 23–34.
22. Sahawneh, S. et al., "Requirements for the Next-Generation Autonomous Vehicle Ecosystem," *IEEE Southeastcon*, Huntsville, AL, 2019.
23. Bloomberg, accessed June 1, 2019, www.bloomberg.com/news/articles/2018-05-21/why-intel-s-mobileye-blew-a-red-light-while-showing-off-its-ride.
24. Schreurs, M.A. and Steuwer, S.D., "Autonomous Driving Political, Legal, Social, and Sustainability Dimensions," in *Autonomous Driving* (Berlin, Heidelberg: Springer, 2016), 149–171.
25. Van Brummelen, J., OBrien, M., Gruyer, D., and Najjaran, H., "Autonomous Vehicle Perception: The Technology of Today and Tomorrow," *Transportation Research Part C: Emerging Technologies* 89 (2018): 384–406.
26. Straub, E.R. and Schaefer, K.E., "It Takes Two to Tango: Automated Vehicles and Human Beings Do the Dance of Driving–Four Social Considerations for Policy," *Transportation Research Part A: Policy and Practice* 122 (2018): 173–183.
27. C. D. of Motor Vehicles, "Testing of Autonomous Vehicles," 2016, accessed June 1, 2019, https://www.dmv.ca.gov/portal/dmv/detail/vr/autonomous/testing.
28. U.S. Department of Transportation, "USDOT Automated Vehicles Activities," 2019, accessed June 1, 2019, https://www.transportation.gov/AV.
29. U.S. Department of Transportation Federal Highway Administration, "Toward Zero Deaths," 2019, accessed June 1, 2019, https://safety.fhwa.dot.gov/tzd.
30. N. C. of State Legislatures, "Autonomous Vehicles Legislative Database," 2017, accessed June 1, 2019, http://www.ncsl.org/research/transportation/autonomous-vehicles-legislative-database.aspx.
31. German Road Traffic Act, "Achtes gesetz zur anderung des straben verkehrsgesetzes," *Bundesgesetzblatt Teil I* 20 (2017): 1648–1650.
32. Mcity, 2018, University of Michigan Mobility Transformation Center, https://mcity.umich.edu/, accessed November 30, 2017.
33. Al-Qadi, I.L., Loulizi, A., Elseifi, M., and Lahouar, S., "The Virginia Smart Road: The Impact of Pavement Instrumentation on Understanding Pavement Performance," *The Journal of Association of Asphalt Paving Technologists* 83 (2004): 427–466.
34. USDOT, 2017, U.S. Department of Transportation Proving Grounds, accessed June 1, 2019, http://www.nationalavpg.com.
35. Favaro, F.M., Nader, N., Eurich, S.O., Tripp, M., and Varadaraju, N., "Examining Accident Reports Involving Autonomous Vehicles in California," *PLoS One* 12 (2017): e0184952.
36. Banerjee, S.S., S. Jha, J. Cyriac, Z. T. Kalbarczyk, and Iyer, R.K., "Hands Off the Wheel in Autonomous Vehicles?: A Systems Perspective on over a Million Miles of Field Data, in *Annual IEEE/IFIP International Conference on Dependable Systems and Networks (DSN)*, Luxembourg City, Luxembourg, 2018, IEEE.
37. ACTS, Advanced Car Technology Systems GmbH & Co, 2018, http://www.acts.de, accessed January 20, 2018.
38. Automotive Enviro Testing, "AET Cold Weather Test Facility & Proving Grounds," 2018, https://www.aettesting.com/, accessed January 20, 2018.
39. ATC, US Army Aberdeen Test Center, 2018, http://www.atc.army.mil, accessed January 20, 2018.
40. ADC, Arctic Driving Center, 2018, http://www.arcticdrivingcenter.fi, accessed January 20, 2018.
41. AF, Arctic Falls Proving Ground, 2018, http://www.arcticfalls.se, accessed January 20, 2018.
42. APL, Arctic Power Lab, 2018, https://www.arcticpower.fi, accessed January 20, 2018.
43. ATS, Arctic Testing Services, 2018, https://www.arctictesting.fi, accessed January 20, 2018.
44. ASTA, "Active Safety Test Area and Zero," 2018, https://www.astazero.com, 2018, accessed January 20, 2018.
45. ATP, Automotive Testing Papenburg GmbH, 2018, https://atppbg.de, accessed January 20, 2018.
46. AARC, Australian Automotive Research Centre, 2018, http://www.aarconline.com, accessed January 20, 2018.
47. ARTC, Automotive Research & Testing Center, 2018, https://www.artc.org.tw, accessed January 20, 2018.
48. BOSCH, "Mobility Solutions Juvincourt," 2018, https://www.bosch-mobility-solutions.com/en/products-and-services/mobility-services/proving-grounds/proving-ground-juvincourt, accessed March 20, 2018.
49. BOSCH, Winter Test Center, 2018, https://www.bosch-mobility-solutions.com/en/products-and-services/mobility-services/proving-grounds/proving-ground-vaitoudden, accessed March 25, 2018.

50. BPG, Bruntingthorpe Proving Ground, 2018, https://www.bruntingthorpe.com, accessed January 20, 2018.
51. Changan Auto & Ford, Proving Ground, 2018, accessed June 1, 2019, http://www.obermeyer-cn.com/en/project/chongqing-testing-grounds-changan.
52. Colmis AB, "Proving Ground," 2018, https://www.colmis.com, accessed March 25, 2018.
53. Cooper, Tire and Vehicle Test Center, 2018, http://coopertire.com/About/Facilities/Technical-and-Test-Centers.aspx, accessed April 21, 2018.
54. FPG, Fowlerville Proving Ground, 2018, http://www.ftt-a.com, accessed April, 21, 2018.
55. GTF, Gerotek Test Facilities, 2018, http://www.gerotek.co.za, accessed April 21, 2018.
56. GCAPS, Global Center for Automotive Performance Simulation, 2018, http://gcaps.net, accessed April 21, 2018.
57. GS Engineering Inc., "Testing and Validation," 2018, http://www.gsengineering.com, accessed April 21, 2018.
58. Icemakers, "Test Tracks," 2018, http://www.icemakers.se, accessed April 17, 2018.
59. Applus+IDIADA, "Test Tracks," 2018, http://www.applusidiada.com, accessed April 17, 2018.
60. CSTB, "Jules Verne Climatic Wind Tunnel," 2018, http://recherche.cstb.fr/en/facilities/climatic-wind, accessed April 17, 2018.
61. Akka Automotive, "Vehicle Testing," 2018, https://www.akka-automotive.com/en/automotive-engineering/validation-launch-management.php, accessed April 17, 2018.
62. Horiba Mira, Testing Center, 2018, https://www.horiba-mira.com, accessed April 17, 2018.
63. MICHELIN, Engineering & Services, 2018, https://www.michelin-engineering-and-services.com, accessed April 17, 2018.
64. FORD, Michigan Proving Ground, 2018, accessed June 1, 2019, https://en.wikipedia.org/wiki/Ford_Proving_Grounds.
65. MTRP, Michigan Technical Resource Park, 2018, https://www.mitrp.com, accessed April 18, 2018.
66. Millbrook, Test World, 2018, https://www.millbrook.co.uk, accessed April 18, 2018.
67. mm-lab Telematics Solutions, Testing Center, 2018, http://www.mmlab.de, accessed April 18, 2018.
68. NATRIP, National Automotive Testing and R&D Infrastructure Project, 2018, http://www.natrip.in, accessed April 18, 2018.
69. NATRAX, National Automotive Testing Track, 2018, http://www.natrip.in/index.php/natrax, accessed April 18, 2018.
70. Porsche, Nardo Technical Center, 2018, http://www.porscheengineering.com/nardo, accessed April 18, 2018.
71. Navistar, Proving Grounds, 2018, http://www.navistarprovinggrounds.com, accessed April 18, 2018.
72. Nurburgring, Proving Ground, 2018, http://www.nuerburgring.de, accessed April 18, 2018.
73. Hodges Transportation, Nevada Automotive Test Center, 2018, http://www.natc-ht.com, accessed April 18, 2018.
74. PMG, Motor Vehicle Test and Research Center, 2018, http://www.pmgtest.com, accessed April 18, 2018.
75. QinetiQ, Hurn Proving Ground, 2018, http://www.qinetiq.com, accessed April 18, 2018.
76. Ride Solutions, Winter Tire Testing, 2018, http://www.ridesolutions.net, accessed April 18, 2018.
77. Smithers, Winter Test Center, 2018, http://www.wintertesting.com, accessed April 18, 2018.
78. SHPG, Southern Hemisphere Proving Grounds, 2018, http://www.shpg.co.nz, accessed April 18, 2018.
79. Millbrook, Test World, 2018, http://www.testworld.fi, accessed April 18, 2018.
80. Thompson, Sub Zero Winter Weather Testing, 2018, http://www.thompsonsubzero.ca, accessed April 18, 2018.
81. TRANSPOLIS, Vehicle Testing, 2018, http://www.transpolis.fr, accessed April 16, 2018.
82. TRC, Transportation Research Center Inc. Proving Ground, 2018, http://www.trcpg.com/, accessed April 16, 2018.
83. UTAC CERAM, Proving Ground, 2018, http://www.utacceram.com, accessed April 16, 2018.
84. Continental, Uvalde Proving Ground, 2018, https://www.continentaltruck.com/truck/company/business-unit/uvalde, accessed April 16, 2018.
85. Wabco, Testing, 2018, http://www.wabco-auto.com, accessed April 16, 2018.
86. Winter Testing Finland, Cold Testing Lapland, 2018, http://www.northernmosttesting.com, accessed April 16, 2018.
87. Ford, Lara/You Yangs Proving Ground, 2018.
88. Yuma, Proving Ground, 2018, www.yuma.army.mil, accessed April 16, 2018.
89. Dixit, V.V., Chand, S., and Nair, D.J., "Autonomous Vehicles: Disengagements, Accidents and Reaction Times, *PLoS One* 11 (2016): e0168054.
90. Janai, J., Guney, F., Behl, A., and Geiger, A., "Computer Vision for Autonomous Vehicles: Problems, Datasets and State-of-the-Art," arXiv preprint arXiv:1704.05519 (2017).

91. Cordts, M., Omran, M., Ramos, S., Rehfeld, T., Enzweiler, M., Benenson, R., Franke, U., Roth, S., and Schiele, B., "The Cityscapes Dataset for Semantic Urban Scene Understanding," in *Proceedings of the IEEE Conference on Computer Vision and Pattern Recognition*, Las Vegas, NV, 2016, 3213–3223.
92. Geiger, A., Lenz, P., and Urtasun, R., "Are We Ready for Autonomous Driving? The Kitti Vision Benchmark Suite, in *Computer Vision and Pattern Recognition (CVPR), 2012 IEEE Conference on*, Providence, RI, 2012, 3354–3361.
93. Nilsson, J., Andersson, P., Gu, I.Y., and Fredriksson, J., "Pedestrian Detection Using Augmented Training Data," in *22nd International Conference on Pattern Recognition (ICPR)*, Stockholm, Sweden, 2014, 4548–4553, IEEE.
94. Waymo, "On the Road to Fully Self-Driving," Waymo Safety Report, 2016, accessed June 1, 2019, https://waymo.com/safety.
95. Dolgov, D., "Google Self-Driving Car Project-Monthly Report-September 2016-on the Road," Technical Report, Google, 2016.
96. Ros, G., Sellart, L., Materzynska, J., Vazquez, D., and Lopez, A.M., "The Synthia Dataset: A Large Collection of Synthetic Images for Semantic Segmentation of Urban Scenes, in *Proceedings of the IEEE Conference on Computer Vision and Pattern Recognition*, Las Vegas, NV, 2016, 3234–3243.
97. Gaidon, A., Wang, Q., Cabon, Y., and Vig, E., "Virtual Worlds as Proxy for Multi-Object Tracking Analysis, in *Proceedings of the IEEE Conference on Computer Vision and Pattern Recognition*, Las Vegas, NV, 2016, 4340–4349.
98. Khastgir, S., Birrell, S.A., Dhadyalla, G., Fulker, D., and Jennings, P.A., "A Drive-In, Driver-in-Loop Simulator for Testing Autonomous Vehicles," in *Driving Simulation Conference Europe*, Tuebingen, Germany, 2015.
99. Baumann, G., Riemer, T., Liedecke, C., Rumbolz, P., Schmidt, A., and Piegsa, A., "The New Driving Simulator of Stuttgart University, in *12th International Stuttgart Symposium*. Stuttgart, Germany, 2012.
100. Suikat, R., "The New Dynamic Driving Simulator at DLR," in *Driving Simulator Conference*, Orlando, FL, 2005, 374–381.
101. Rentrope, J., Midence, M., Akbas, M.I., and Razdan, R., "Verification of Autonomous Vehicles Using Simulations with String Based Scenarios," in *Institute of Industrial & System Engineers (IISE) Conference and Expo*, Orlando, FL, 2018.
102. Richter, S.R., Vineet, V., Roth, S., and Koltun, V., "Playing for Data: Ground Truth from Computer Games," in *European Conference on Computer Vision*, Amsterdam, The Netherlands, 2016, Springer, 102–118.
103. Zhao, D., Huang, X., Peng, H., Lam, H., and LeBlanc, D.J., "Accelerated Evaluation of Automated Vehicles in Car-Following Maneuvers," *IEEE Transactions on Intelligent Transportation Systems* 19 (2018): 733–744.
104. Casas, J., Ferrer, J.L., Garcia, D., Perarnau, J., and Torday, A., "Traffic Simulation with Aimsun," in Barceló, J. (ed.) *Fundamentals of Traffic Simulation*, International Series in Operations Research & Management Science, Vol. 145 (New York: Springer, 2010): 173–232.
105. Shah, S. et al., "AirSim: High-Fidelity Visual and Physical Simulation for Autonomous Vehicles," in *11th Conference on Field and Service Robotics (FSR)*, Zurich, Switzerland, 2017.
106. Apollo Auto, The Apollo Open Platform, 2018, http://apollo.auto, accessed July 30, 2018.
107. Lin, Y. and Eskandarian, A., "Experimental Evaluation of Cooperative Adaptive Cruise Control with Autonomous Mobile Robots, in *Conference on Control Technology and Applications (CCTA), IEEE*, Kohala Coast, Hawaii, 2017, 281–286.
108. Best, A., Narang, S., Barber, D., and Manocha, D., "Autonovi: Autonomous Vehicle Planning with Dynamic Maneuvers and Traffic Constraints, in *Intelligent Robots and Systems (IROS), 2017 IEEE/RSJ International Conference on, IEEE*, Vancouver, BC, Canada, 2629–2636.
109. Kaminski, T., Niezgoda, M., Gacparska, J., Grzeszczyk, R., and Filipek, P., "Advanced Car Driving Simulator - As 1200-6, *Journal of KONES* 18 (2011): 173–178.
110. Cesium, "Autonomous Car Simulation," 2018, https://cesium.com, accessed July 30, 2018.
111. monoDrive, "Autonomous Vehicle Simulator," 2018, https://monodrive.io, accessed July 30, 2018.
112. Blender Online Community, "Blender - A 3D Modeling and Rendering Package," Blender Foundation, Blender Institute, Amsterdam, 2018.
113. Naujoks, F., Wiedemann, K., Schomig, N., Jarosch, O., and Gold, C., "Expert Based Controllability Assessment of Control Transitions from Automated to Manual Driving," *MethodsX* 5 (2018): 579–592.
114. Dosovitskiy, A., Ros, G., Codevilla, F., Lopez, A., and Koltun, V., "CARLA: An Open Urban Driving Simulator," in *Proceedings of the 1st Annual Conference on Robot Learning*, Mountain View, CA, 2017, 1–16.
115. Cognata, Cognata, Deep Learning Autonomous Simulation, 2018, http://cognata.com, accessed July 30, 2018.

116. Michalaka, D. and Yin, Y., "Managed Lanes Operations and Simulation Using Corsim, IRF, 2013, 19.

117. NVIDIA, "Drive Constellation, Virtual Reality Autonomous Vehicle Simulator," 2018, https://www.nvidia.com/en-us/self-driving-cars/drive-constellation, accessed July 28, 2018.

118. Gomes, P., Olaverri-Monreal, C., Ferreira, M., and Damas, L., "Driver-Centric Vanet Simulation, in T. Strang, A. Festag, A. Vinel, R. Mehmood, C. Rico Garcia, and M. Rockl (Eds.), *Communication Technologies for Vehicles* (Berlin, Heidelberg: Springer, 2011), 143–154.

119. Papadimitriou, S. and Choudhury, C.F., "Transferability of Car-Following Models between Driving Simulator and Field Traffic," *Transportation Research Record: Journal of the Transportation Research Board* (2017) 60–72.

120. rfPro, Driving Simulation, 2018, http://www.rfpro.com/driving-simulation, accessed July 28, 2018.

121. dSPACE GmbH, Simulation Testing, 2018.

122. Paull, L., Tani, J., Ahn, H., Alonso-Mora, J., Carlone, L., Cap, M., Chen, Y.F., Choi, C., Dusek, J., Fang, Y. et al., "Duckietown: An Open, in Expensive and Flexible Platform for Autonomy Education and Research," *Robotics and Automation (ICRA), 2017 IEEE International Conference on*, IEEE, 1497–1504.

123. Koenig, N. and Hsu, J., The Many Faces of Simulation: Use Cases for a General Purpose Simulator, in *Proc. of the ICRA*, Vol. 13, Karlsruhe, Germany, 2013, 10–11.

124. Hou, Y., Zhao, Y., Wagh, A., Zhang, L., Qiao, C., Hulme, K.F., Wu, C., Sadek, A.W., and Liu, X., "Simulation-Based Testing and Evaluation Tools for Transportation Cyber–Physical Systems," *IEEE Transactions on Vehicular Technology* 65 (2016): 1098–1108.

125. The MathWorks, Inc., "MATLAB and Statistics Toolbox Release R2018a," 2018.

126. Horni, A., Nagel, K., Axhausen, K.W., *The Multi-Agent Transport Simulation MATSim* (London: Ubiquity Press, 2016).

127. metamoto, "Simulation as a Service for Automated Vehicles," 2018, https://www.metamoto.com, accessed July 30, 2018.

128. Son, J., Reimer, B., Mehler, B., Pohlmeyer, A., Godfrey, K., Orszulak, J., Long, J., Kim, M., Lee, Y., and Coughlin, J.F., "Age and Cross-Cultural Comparison of Drivers Cognitive Workload and Performance in Simulated Urban Driving," *International Journal of Automotive Technology* 11 (2010): 533–539.

129. Echeverria, G. et al., "Simulating Complex Robotic Scenarios with MORSE," in Noda, I., Ando, N., Brugali, D., Kuffner, J.J. (eds.) *Simulation, Modeling, and Programming for Autonomous Robots. SIMPAR 2012*, Lecture Notes in Computer Science, Vol. 7628 (Berlin, Heidelberg: Springer, 2012).

130. Wu, Y., Boyle, L.N., McGehee, D., Roe, C.A., Ebe, K., and Foley, J., "Modeling Types of Pedal Applications Using a Driving Simulator," *Human Factors* 57 (2015): 1276–1288.

131. Dupuis, M., Strobl, M., and Grezlikowski, H., "Opendrive 2010 and Beyond – Status and Future of the De Facto Standard for the Description of Road Networks," in *Proc. Driving Simulation Conference DSC Europe*, Paris, S, 231–242.

132. Nakasone, A., Prendinger, H., Miska, M., Lindner, M., Horiguchi, R., and Kuwahara, M., "OpenEnergySim: A 3D Internet Based Experimental Framework for Integrating Traffic Simulation and Multi-User Immersive Driving," in *Proceedings of the International ICST Conference on Simulation Tools and Techniques*, Barcelona, Spain, 2011, Institute for Computer Sciences, Social Informatics and Telecommunications Engineering, 490–498.

133. OSU, Driving Simulation Laboratory at The Ohio State University, 2018, http://drivesim.osu.edu/facility, accessed July 30, 2018.

134. Cameron, G.D. and Duncan, G.I., "Paramicsparallel Microscopic Simulation of Road Traffic," *The Journal of Supercomputing* 10 (1996): 25–53.

135. Vissim, P., "7.0 User Manual," PTV AG, Karlsruhe, Niemcy, 2014.

136. Righthook, RightHook Simulation, RightMap, RightWorld, 2018, https://righthook.io, accessed July 30, 2018.

137. Zhang, C., Liu, Y., Zhao, D., and Su, Y., "Roadview: A Traffic Scene Simulator for Autonomous Vehicle Simulation Testing, in *Intelligent Transportation Systems (ITSC), 2014 IEEE 17th International Conference on*, Qingdao, China, 2014, IEEE, 1160–1165.

138. Jurecki, R., Poliak, M., and Jaskiewicz, M., "Young Adult Drivers: Simulated Behaviour in a Car-Following Situation," *Promet-Traffic & Transportation* 29 (2017): 381–390.

139. Udacity, Udacity Self-Driving Car Simulator, 2018, https://github.com/udacity/self-driving-car-sim, accessed July 28, 2018.

140. Stratulat, A.M., Roussarie, V., Vercher, J.-L., and Bourdin, C., "Does Tilt/Translation Ratio Affect Perception of Deceleration in Driving Simulators?," *Journal of Vestibular Research* 21 (2011): 127–139.

141. Samuel, S., Borowsky, A., Zilberstein, S., and Fisher, D.L., "Minimum Time to Situation Awareness in Scenarios Involving Transfer of Control from an Automated Driving Suite," *Transportation Research*

Record: Journal of the Transportation Research Board 2602 (2016): 115–120.

142. Morris, K.L., Fabiano, G.A., and Hulme, K.F., "Leveraging Simulation to Augment Risky Driving Attitudes and Behaviors," in *The Interservice/Industry Training, Simulation and Education Conference (I/ITSEC)*, Orlando, FL, 2014.
143. Johnson-Roberson, M., Barto, C., Mehta, R., Sridhar, S.N., Rosaen, K., and Vasudevan, R., "Driving in the Matrix: Can Virtual Worlds Replace Human-Generated Annotations for Real World Tasks?," in *IEEE International Conference on Robotics and Automation*, Stockholm, Sweden, 2016, 1–8.
144. Vaughan, R., "Massively Multi-Robot Simulation in Stage," *Swarm Intelligence* 2 (2008): 189–208.
145. Russell, H.E., Harbott, L.K., Nisky, I., Pan, S., Okamura, A.M., and Gerdes, J.C., "Motor Learning Affects Car-to-Driver Handover in Automated Vehicles," *Trials* 6 (2016): 6.
146. STISIM, STISIM Drive DT Software, 2018, https://stisimdrive.com, accessed July 28, 2018.
147. Krajzewicz, D., Erdmann, J., Behrisch, M., and Bieker, L., "Recent Development and Applications of SUMO - Simulation of Urban MObility," *International Journal on Advances in Systems and Measurements* 5 (2012): 128–138.
148. Schiavullo, R., "SYNCITY Virtual Hyper Realistic Simulator for Machine Learning to Train ADAS Systems," *Virtual Reality* 4 (2018): 06.
149. TASS International, PreScan, 2018.
150. Toyota, Toyota Develops World-Class Driving Simulator, November 26, 2007, http://www.toyota.co.jp/en/news/07/1126-1.html, accessed December 28, 2017.
151. C. C. A. Carnegie Mellon University, General Motors, Darpa Urban Challenge Final Report for Tartan Racing, 2007.
152. Uber, "Engineering Ubers Self-Driving Car Visualization Platform for the Web," 2017, https://eng.uber.com/atg-dataviz/, accessed: March 28, 2018.
153. Carpin, S., Lewis, M., Wang, J., Balakirsky, S., and Scrapper, C., "Usarsim: A Robot Simulator for Research and Education," in *Robotics and Automation, 2007 IEEE International Conference on*, IEEE, 1400–1405.
154. Rohmer, E., Singh, S.P.N., and Freese, M., "V-Rep: A Versatile and Scalable Robot Simulation Framework, in *Proc. of the International Conference on Intelligent Robots and Systems (IROS)*, Tokyo, Japan, 2013.
155. Grant, P., Artz, B., Greenberg, J., and Cathey, L., "Motion Characteristics of the Virttex Motion System," in *Proceedings of the 1st Human-Centered Transportation Simulation Conference*, Iowa City, IO, 2001, 4–7.
156. Schoner, H.-P., "Simulation in Development and Testing of Autonomous Vehicles," in *18. Internationales Stuttgarter Symposium*, Stuttgart, Germany, 2018, Springer, 1083–1095.
157. Oeltze, K. and Schießl, C., "Benefits and Challenges of Multi-Driver Simulator Studies, *IET Intelligent Transport Systems* 9 (2015): 618–625.
158. Jansson, J., Sandin, J., Augusto, B., Fischer, M., Blissing, B., and Kallgren, L., "Design and Performance of the VTI SIM IV, in *Driving Simulation Conference*, Paris, France, 2014, 128–138.
159. Noyce, D.A. and Chitturi, M.V., "Virtual Road Safety Audits: Recommended Procedures for Using Driving Simulation and Technology to Expand Existing Practices," 2019, [online] accessed June 6, 2019, https://rosap.ntl.bts.gov/view/dot/35567.
160. Li, L., Huang, W.-L., Liu, Y., Zheng, N.-N., Wang, F.-Y., "Intelligence Testing for Autonomous Vehicles: A New Approach," *IEEE Transactions on Intelligent Vehicles* 1 (2016): 158–166.
161. Winner, H., Lemmer, K., Form, T., and Mazzega, J., "PEGASUSFirst Steps for the Safe Introduction of Automated Driving," in *Automated Vehicles Symposium*, San Francisco, CA, Springer, 185–195.
162. Hallerbach, S., Xia, Y., Eberle, U., and Koester, F., "Simulation-Based Identification of Critical Scenarios for Cooperative and Automated Vehicles," Technical Report, SAE Technical Paper, 2018.
163. Editor, N., "A Match Made in Chip Verification Heaven: Simulation and Emulation," 2019, [online] T&VS, accessed June 6, 2019, https://www.testandverification.com/thought-leadership/a-match-made-in-chip-verification-heaven-simulation-and-emulation-21-september-2016/.
164. Cambria, E. and White, B., "Jumping NLP Curves: A Review of Natural Language Processing Research," *IEEE Computational Intelligence Magazine* 9 (2014): 48–57.

Contact Information

EDGEresearch@sae.org.

The authors of this document together with the SAE Team responsible for its creation join in expressing our deepest appreciation to all of the individuals who contributed.

www.ingramcontent.com/pod-product-compliance
Lightning Source LLC
LaVergne TN
LVHW070940210926
848482LV00003B/269